Computational Approaches in Molecular Interaction Analysis

Predicting Protein Complexes, Host-Pathogen Interactions, and Protein Function from Interactome Data

Computational Approaches in Molecular Interaction Analysis

Predicting Protein Complexes, Host-Pathogen Interactions, and Protein Function from Interactome Data

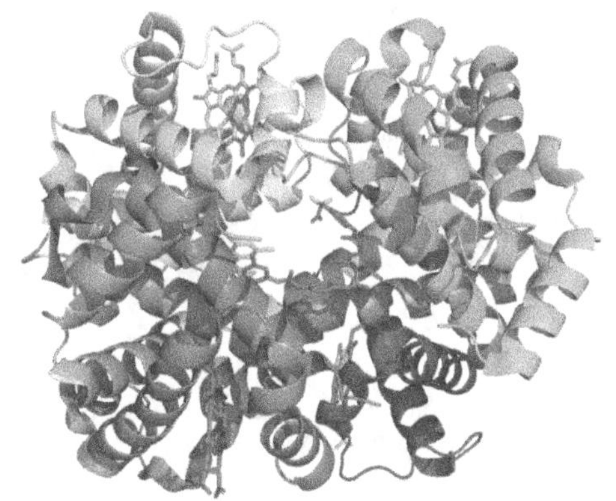

Edited by

Limsoon Wong

National University of Singapore, Singapore

NEW JERSEY · LONDON · SINGAPORE · GENEVA · BEIJING · SHANGHAI · TAIPEI · CHENNAI

Published by

World Scientific Publishing Co. Pte. Ltd.

5 Toh Tuck Link, Singapore 596224

USA office: 27 Warren Street, Suite 401-402, Hackensack, NJ 07601

UK office: 57 Shelton Street, Covent Garden, London WC2H 9HE

British Library Cataloguing-in-Publication Data
A catalogue record for this book is available from the British Library.

COMPUTATIONAL APPROACHES IN MOLECULAR INTERACTION ANALYSIS
Predicting Protein Complexes, Host-Pathogen Interactions, and
Protein Function from Interactome Data

ISBN 978-981-98-0717-8 (hardcover)
ISBN 978-981-98-0718-5 (ebook for institutions)
ISBN 978-981-98-0719-2 (ebook for individuals)

For any available supplementary material, please visit
https://www.worldscientific.com/worldscibooks/10.1142/14157#t=suppl

Typeset by Stallion Press
Email: enquiries@stallionpress.com

Introduction

The landscape of molecular biology has been profoundly transformed by the advent of high-throughput technologies and computational methods, which have ushered in an era of unprecedented data generation and analysis capabilities. Among the many areas that have benefited from these advancements are protein complex prediction, host–pathogen interaction (HPI) prediction and protein function prediction from protein–protein interactions (PPIs). This book brings together a collection of research articles — published over the past few years in the *Journal of Bioinformatics and Computational Biology* — that address significant challenges and present innovative solutions in these fields.

Part I: Protein Complex Prediction

In the first part of the book, we delve into the dynamic and intricate nature of protein interactions and protein complexes. Despite the static nature of PPI networks obtained through current technologies; proteins operate in a highly dynamic fashion.

Chapter 1 opens with an exploration of the challenges in deriving dynamic protein complexes from static interaction data. Yong and Wong[1] identify three primary difficulties: the overlapping nature of many protein complexes, the sparsity of interactions for many protein complexes, and the sensitivity of small complexes to extraneous edges in a PPI network. Their insights pave the way for improving complex discovery algorithms.

In **Chapter 2**, Yao *et al.*[2] present a novel approach to de-noising PPI networks using a variational graph auto-encoder. By reconstructing reliable PPI networks, they significantly enhance the accuracy of protein complex detection across various datasets, demonstrating the potential of integrating advanced machine learning techniques with biological data.

Chapter 3 introduces the gene expression and core-attachment (GECA) method by Noori *et al.*[3] GECA combines common neighbour techniques with biological information to identify core proteins and enhances the attachment technique by including proteins with high gene expression similarity but low closeness. GECA successfully identifies biologically significant protein complexes, highlighting the importance of integrating gene expression profiles into PPI analysis.

Chapter 4 concludes this part with the IPC-RPIN method of Zhang *et al.*,[4] which represents a significant advancement in protein complex prediction by integrating network topology, gene expression data and GO annotation information. Recognising that high-throughput data are often incomplete and contain spurious interactions, IPC-RPIN reconstructs PPI networks by refining the interactions using a combination of topological features, gene expression profiles, and GO annotations. The method not only addresses the reliability issues of PPI networks but also improves the accuracy of protein complex prediction. Importantly, IPC-RPIN excels at identifying small protein complexes, which are often neglected by other methods, thereby providing a more comprehensive understanding of the cellular machinery.

Part II: Host–Pathogen Interaction Prediction

The second part of the book shifts focus to the critical area of HPIs, which are essential for understanding infectious diseases and developing therapeutic interventions.

Chapter 5 provides a comprehensive review by Zhou *et al.*[5] of computational studies on HPIs, focusing on methodologies, tools, and advancements in understanding infectious diseases through computational biology. They explore predictive models for PPIs between hosts and pathogens, discuss network analysis techniques to decipher interaction patterns, and highlight structural insights gained from computational modeling. The chapter also surveys databases and software tools essential for data integration and analysis in HPI research. It concludes by outlining current challenges, such as data accuracy and biological relevance, and proposes future directions to enhance predictive models considering dynamic host-pathogen dynamics and evolutionary factors, making it an indispensable resource for researchers in the fields of computational biology and infectious disease studies.

In **Chapter 6**, Basit *et al.*[6] tackle the challenge of developing machine learning models to predict HPIs by exploring methodologies for selecting negative training examples, the impact of sample weighting on prediction accuracy, and the optimal ratio of positive to negative samples in training datasets. This chapter serves as a useful guide for researchers and practitioners interested in leveraging computational approaches to study HPIs and underscores the importance of machine learning in infectious disease research.

Chapter 7 presents an improved method by Kim *et al.*[7] for predicting virus–human protein interactions. They introduce an innovative method using support vector machines (SVMs) that integrates key features of viruses and human proteins, including amino acid triplet frequencies and composition differences. Their approach achieves superior performance compared to existing methods. By incorporating gene ontology (GO) annotations, the approach also predicts novel virus–human interactions, underscoring its potential for uncovering new pathogenic mechanisms and therapeutic targets.

Chapter 8 addresses performance evaluation issues in HPI prediction. Abbasi and Minhas[8] critique commonly used validation techniques and propose more effective metrics, emphasising the need for realistic evaluation frameworks to enhance predictive models' reliability.

Part III: Protein Function Prediction from Protein–Protein Interactions

The final part of the book explores methodologies for predicting protein functions based on PPI networks, an area crucial for understanding cellular processes and disease mechanisms.

Chapter 9 introduces a multi-label protein function prediction method by Saha *et al.*[9] This approach leverages GO-based neighbourhood analysis and physico-chemical features, demonstrating significant improvements in functional group assignment for both human and yeast proteins.

In **Chapter 10**, Saha *et al.*[10] extend their work by incorporating dynamic PPI networks, deduced using gene expression data, to predict protein functions. Their methodology accounts for the temporal nature of protein interactions, providing a more accurate functional annotation.

Chapter 11 concludes with an iterative model by Wang and Hou[11] for protein function prediction from a PPI network. Their iterative approach addresses limitations in previous prediction algorithms by integrating interactions involving unannotated proteins into the prediction process. By iteratively refining predictions based on interaction information and network topology, their approach shows substantial performance enhancements, particularly when unannotated proteins constitute a significant portion of the network.

Conclusion

This collection represents a broad spectrum of innovative research and methodologies that address fundamental challenges in protein complex prediction, HPI prediction and protein function prediction. By bringing these diverse approaches together, this book aims to provide a comprehensive resource for researchers and practitioners in the field of computational biology, fostering further advancements and applications in understanding cellular mechanisms and disease processes.

Limsoon Wong
13 June 2024

References

1. Yong CH, Wong L, From the static interactome to dynamic protein complexes: Three challenges, *J Bioinform Comput Biol* **13**:1571001, 2015.
2. Yao H, Guan J, Liu T, Denoising protein–protein interaction network via variational graph auto-encoder for protein complex detection, *J Bioinform Comput Biol* **18**:2040010, 2020.
3. Noori S, Al-A'Araji N, Al-Shamery E, Identifying protein complexes from protein–protein interaction networks based on the gene expression profile and core-attachment approach, *J Bioinform Comput Biol* **19**:2150009, 2021.
4. Zhang W, Xu J, Li Y, Zou X, Integrating network topology, gene expression data and GO annotation information for protein complex prediction, *J Bioinform Comput Biol* **17**:1950001, 2019.
5. Zhou H, Jin J, Wong L, Progress in computational studies of host-pathogen interactions, *J Bioinform Comput Biol* **11**:1230001, 2013.
6. Basit AH, Abbasi WA, Asif A, Gull S, Minhas FUAA, Training host–pathogen protein–protein interaction predictors, *J Bioinform Comput Biol* **16**:1850014, 2018.

7. Kim B, Alguwaizani S, Zhou X, Huang DS, Park B, Han K, An improved method for predicting interactions between virus and human proteins, *J Bioinform Comput Biol* **15**:1650024, 2017.

8. Abbasi WA, Minhas FUAA, Issues in performance evaluation for host–pathogen protein interaction prediction, *J Bioinform Comput Biol* **14**:1650011, 2016.

9. Saha S, Prasad A, Chatterjee P, Basu S, Nasipuri M, Protein function prediction from protein–protein interaction network using gene ontology-based neighborhood analysis and physico-chemical features, *J Bioinform Comput Biol* **16**:1850025, 2018.

10. Saha S, Prasad A, Chatterjee P, Basu S, Nasipuri M, Protein function prediction from dynamic protein interaction network using gene expression data, *J Bioinform Comput Biol* **17**:1950025, 2019.

11. Wang D, Hou J, Explore the hidden treasure in protein–protein interaction networks – an iterative model for predicting protein functions, *J Bioinform Comput Biol* **13**:1550026, 2015.

Contents

Part I
Protein Complex Prediction

CHAPTER 1

From the Static Interactome to Dynamic Protein Complexes: Three Challenges[a]

Chern Han Yong[*,‡] and Limsoon Wong[†,§]

*Graduate School for Integrative Sciences and Engineering
National University of Singapore
28 Medical Drive, Singapore 117456

†School of Computing, National University of Singapore
13 Computing Drive, Singapore 117417
‡cherny@nus.edu.sg
§wongls@comp.nus.edu.sg

Protein interactions and complexes behave in a dynamic fashion, but this dynamism is not captured by interaction screening technologies, and not preserved in protein–protein interaction (PPI) networks. The analysis of static interaction data to derive dynamic protein complexes leads to several challenges, of which we identify three. First, many proteins participate in multiple complexes, leading to overlapping complexes embedded within highly-connected regions of the PPI network. This makes it difficult to accurately delimit the boundaries of such complexes. Second, many condition- and location-specific PPIs are not detected, leading to sparsely-connected complexes that cannot be picked out by clustering algorithms. Third, the majority of complexes are small complexes (made up of two or three proteins), which are extra sensitive to the effects of extraneous edges and missing co-complex edges. We show that many existing complex-discovery algorithms have trouble predicting such complexes, and show that our insight into the disparity

---

[a] This article was previously published in *Journal of Bioinformatics and Computational Biology*. Vol: 13, No. 2 (2015), 1571001 (25 pages).

[‡] Corresponding author.

between the static interactome and dynamic protein complexes can be used to improve the performance of complex discovery.

Keywords: Protein complex; protein interaction; dynamism.

1. Introduction

In the cell, many proteins bind physically to form stoichiometrically-stable multiprotein structures called protein complexes. Protein complexes perform a wide variety of molecular functions in many cellular processes, thus it is important to determine the set of complexes in the cell to gain an understanding of the mechanism, organization, and regulation of these processes. Since proteins in a complex interact physically, many algorithms have been proposed to analyze protein–protein interaction (PPI) data to discover protein complexes.

The general strategy underlying most complex-discovery algorithms is to represent PPI data as a PPI network (PPIN), where vertices represent proteins and edges represent interactions between proteins, and then find clusters of highly interconnected proteins within the PPIN as protein complexes. Over the past decade, these algorithms have grown in sophistication and variety, and have incorporated increasing amounts of useful biological insights in their designs. However, the performance of most of these approaches, even under optimal conditions, still leaves room for improvement: For example, even in yeast with decently-comprehensive PPI data, accurate prediction of complexes at fine resolution remains difficult.

One main stumbling block is that the representations and analyses of PPIs for the purpose of complex prediction have been overwhelmingly static, even though it has been well understood that proteins and complexes exhibit a sophisticated dynamism in behavior. Proteins interact in a dynamic fashion, with a variety of interaction timings, locations, and affinities. These are mediated by a wide range of factors including cellular state, cellular processes, and the interaction environment.[1] Correspondingly, protein complexes exhibit dynamic behavior which are in fact important functional mechanisms, for example to allow complexes to be formed only at certain times, or to vary the composition of complexes to modulate or activate their functions. However, due to limitations in PPI-detection methodologies, it is difficult to interrogate the dynamics of PPIs (i.e. when, where, and how a protein interacts with others). Furthermore, this dynamism also precludes a faithful interrogation of PPIs in the cell (e.g. condition-specific PPIs may be missed, or spurious PPIs may be detected in non-physiological experimental systems). Moreover, the representation of PPIs in the PPIN does not preserve any information about the dynamics of PPIs. Thus there exists a disparity between the dynamic nature of PPIs and protein complexes on one hand, and the static representation and analysis of the PPIN on the other hand.

We identify three challenges in protein-complex discovery that arise from, or are exacerbated by, this static view of PPIs and protein complexes. First, many complexes are embedded within highly-connected regions of the PPIN, with many extraneous edges connecting a complex's member proteins to other proteins outside the complex. This arises because many proteins participate in multiple distinct complexes, resulting in complexes overlapping each other in dense regions in the PPIN. Spuriously-detected interactions further contribute to this problem. Second, many complexes exist in sparse regions of the network, so that proteins within the complexes are not densely interconnected. This arises from undetected condition-specific, location-specific, or transient PPIs. Third, many complexes are small (that is, composed of two or three proteins), making measures of important topological features, such as density, ineffectual. This is further exacerbated by extraneous or missing interactions which can embed a small complex in a larger clique, or disconnect it entirely.

In this paper, we evaluate the performance of various complex-discovery algorithms, covering different types of approaches, in the prediction of yeast and human complexes. In particular, we highlight the unsatisfactory performance in predicting complexes within highly-connected regions, complexes within sparse regions, and small complexes, and discuss how an understanding of the dynamics of protein interactions may be used to address the shortcomings of these algorithms with respect to these specific challenges.

A number of surveys on complex discovery have been published in recent years. Li *et al.*[2] in 2010 surveyed a number of complex-discovery algorithms, and categorized them according to the types of data used and the features of the algorithms. Srihari and Leong[3] in 2013 further showed that complex-discovery algorithms have evolved to incorporate increasing amounts of biological information in their designs, leading to improved performance and new biological insights. Most recently, Chen *et al.*[4] also surveyed and categorized various complex-discovery algorithms, with a distinct category for algorithms that explicitly model the dynamism of PPIs. Since descriptions and taxonomies of complex-discovery algorithms are already covered in these surveys, our paper instead emphasizes specific challenges raised by the dynamism of PPIs, and evaluates a few classic and recent algorithms with respect to these challenges.

In Sec. 2, we elaborate on protein interactions and protein complexes in the cell, with an emphasis on the dynamism of their behaviors. We give a brief background on PPI-screening technologies and their inadequacies, particularly in capturing such dynamism. In Sec. 3, we show how the three challenges in complex discovery follow from the analysis of static PPIs. In Sec. 4, we describe our experiments to evaluate five clustering algorithms in yeast and human complex discovery, with an emphasis on their shortcomings with respect to the three challenges that we have highlighted. In Sec. 5, we conclude our findings, and describe some approaches that help to address these challenges, although much room remains for improvement.

2. Background: From Interactome to Complexome

In the study of protein complexes, the interactome refers to the set of cellular physical PPIs, while the complexome describes the set of cellular complexes. Since complexes consist of physically-interacting proteins, they correspond to groups of proteins with high degrees of co-interaction in the interactome. Thus, deriving the complexome from the interactome is a fruitful strategy that has been well researched over the past decade. Many challenges have been acknowledged in this strategy, a significant portion of which we distil as the 'disparity' between the static interactome and the complexome: Due to limitations in detection technologies and methodologies (which have only recently begun to be surpassed), the views and analyses of the interactome and complexome have been overwhelmingly static, without consideration of the dynamic nature of PPIs and the corresponding dynamism of protein complexes.

2.1. *Dynamism of Protein Interactions*

In fact, the static interactome, understood as the set of PPIs that exist in a cell, is a mere shadow of the dynamic and complex lives of PPIs in reality, which involve a wide range of interaction timings, locations, and binding affinities.

A protein with multiple interaction partners does not necessarily interact with all of them simultaneously. When a protein interacts, and which partner it interacts with, are controlled by different cellular mechanisms. For example, co-localization of the interactors in time and space, as well as the local concentration of the interactors, are controlled by expression, mRNA degradation, protein transport, protein secretion, protein degradation; the binding affinities of different interactors are controlled through post-translational modification of the interactors, or changes to the physiochemical environment, for example by the concentration of effector molecules like ATP that may change binding affinity.[1]

Different classes of PPI binding affinities have been proposed[1,5,6]: permanent interactions, with the strongest binding affinity, are irreversible; weak transient interactions, with the weakest binding affinity, are reversible, and involve proteins that switch between both bound and unbound states *in vivo*; strong transient interactions lie between permanent interactions and weak transient interactions, and are reversible when triggered, for example by ligand binding. PPIs can also be characterized as obligate or non-obligate: proteins with obligate interactions cannot exist as stable structures on their own, and are frequently bound to their partners upon translation and folding; conversely, proteins with non-obligate interactions can exist as stable structures both in bound and unbound states.

A study of protein hubs (proteins with a large number of interaction partners) with gene-expression data has led to a proposed distinction between date hubs and party hubs[7,8]: party hubs interact with all of their partners simultaneously as a large complex, while date hubs interact with its partners in mutually-exclusive times, and are believed to link diverse biological processes together in the PPIN.

2.2. *Dynamism of Protein Complexes*

Consequently, complexes display a range of dynamism in their formation, composition, and stability, which impart important functional mechanisms to the complexes' activities. For example, the highly conserved Cdc28p (a cyclin-dependent kinase or CDK) yeast protein regulates the cell-cycle by forming complexes with different cyclin proteins that phosphorylate different substrates to promote entry into different cell-cycle phases[9,10]: progressing through the cell-cycle phases, these include Cdc28p forming complexes with Cln3p to enter the cycle, with Cln1,2p in G_1 phase, with Clb5,6p to begin replication in S phase, and with Clb1,2,3,4p to enter M phase (see Fig. 7(a)). These complexes are themselves regulated through binding with cyclin-dependent kinase inhibitors (CKIs) such as Sic1p.

An integrated analysis of protein complexes with cell-cycle expression data revealed "just-in-time" assembly of most cell-cycle-related complexes in yeast[11]: some subunits of complexes are constitutively expressed (static proteins), while other subunits are expressed only when needed (dynamic proteins), so that the entire complex can be assembled only in specific cell-cycle phases without having to transcriptionally regulate all the subunits of the complex.

The dynamism of complexes also gives them a modular architecture in function and composition, which has been described with the core-attachment model of complexes.[12] Here, the core of a complex consists of proteins that interact permanently, while attachment proteins are recruited to the core via less permanent interactions, which may modulate or activate the function of the complex.

2.3. *Interactome Screening Technologies*

The dynamism of PPIs, which provides such important functional mechanisms for complexes, is not captured in the static interactome. A chief reason for this is the technological limitations of past high-throughput PPI screening experiments, which has only recently begun to be surpassed.

In the past decade, the two commonly used methods for high-throughput screening of PPIs are based on the yeast two-hybrid assay (Y2H), which detects binary interactions, and the tandem affinity purification with mass spectrometry (TAP-MS) method, which detects co-complex interactions. The Y2H method uses a fragmented transcription factor to detect the interaction between a bait protein and a prey protein: when the proteins interact, they cause the transcription of a reporter gene.[13] A recent survey of advances in Y2H technology is provided by Bruckner *et al.*[14]

The Y2H assay is able to detect transient or weak interactions, but is limited to only direct physical PPIs: interactions between co-complex proteins (proteins in the same complex) that do not physically interact with each other are not detected. Y2H assays interactions at non-physiological conditions (e.g. the bait and prey proteins may be overexpressed, co-expressed, or post-translationally modified, whereas they may not be *in vivo*), so some interactions may be spuriously detected. Since interactions are interrogated in a controlled homogeneous cellular state, those that occur in other condition-specific states (such as different cell-cycle or perturbation states) may not be captured. Furthermore, some interacting proteins are unable to localize in the nucleus, or cannot interact in the nucleus' environment, so these interactions are not detected. Conversely, proteins that never co-localize *in vivo* and are thus unable to interact might be wrongly detected as interacting in the nucleus.

Aside from the above problems, Y2H also suffers from the variability inherent in interrogating biological systems, leading to poor reproducibility across multiple screens.

TAP-MS[15] allows a bait protein to complex with other proteins under physiological conditions, and washes it through affinity columns to detect its co-complex proteins (the prey proteins) via mass spectrometry. A survey of recent advances in MS-based methods is provided by Gavin *et al.*[16]

TAP-MS typically only captures strong interactions. Unlike the Y2H assay, TAP-MS retrieves proteins co-complexed with the bait protein, including those that are only indirectly-associated via bridging proteins. Furthermore, for bait proteins that form multiple distinct complexes, all the proteins that form the union of these complexes may be purified and detected. To uncover the PPIs from the purified complexes, further processing is needed,[12,17] though this may still lead to false positives (direct interactions imputed between indirectly-associated proteins) and false negatives (interactions between prey proteins not imputed).

In many TAP-MS assays, the bait protein is expressed by non-natural promoters, leading to its over-expression over physiological levels[16] (although in some studies its expression is controlled by natural promoters.[12,18]).

Under TAP-MS, protein complexes in any subcellular location can be purified. Furthermore, since a heterogeneous collection of cells are purified, complexes present in multiple cellular conditions, such as various cell-cycle and growth states, may be retrieved.[12,18] Nevertheless, complexes present only in other conditions, such as specific perturbation states, are not retrieved. Only recently have researchers begun interrogating the composition of complexes under different perturbation states, for example with affinity purification with selected reaction monitoring,[19] or affinity purification combined with sequential window acquisition of all theoretical spectra.[20] Both works represent key advances in methodologies that will allow dynamic and condition-specific views of interactomes in the near future; but for now, the range of the proteins and PPIs probed, as well as the conditions tested, remain limited.

2.4. *The Static Interactome*

As described above, the Y2H and TAP-MS methods do not capture timing (i.e. simultaneity) or localization information about the PPIs. For interactions whose affinities are dependent on molecular trigger events such as phosphorylation, information about such molecular triggers is lost, and moreover interactions whose triggers are not activated are not captured. Neither Y2H nor TAP-MS interrogate interactions with respect to cellular states: Under Y2H, interactions are assayed in a homogeneous cellular state which is frequently non-physiological; while under TAP-MS, interactions are frequently interrogated in heterogeneous cellular growth states, so that proteins present in complexes from various growth states are retrieved as an undifferentiated set. Moreover, complexes present only in specific perturbation conditions, which are absent from the cells, are not found. The PPIs obtained thus represent a static interactome, lacking the dynamism that imparts important functional mechanisms to the PPIs and the complexes that they comprise; at the same time, this dynamism precludes an accurate screening of the interactome.

The interactome is frequently represented as a PPIN, with vertices representing proteins and edges representing interactions. This representation itself is a simplification of the cellular organization of PPIs: Aside from missing information about interaction timing, location, affinity, and cellular state, the representation of each protein as a single vertex conflates the multiple copies of each protein that exist in the cell into a single entity: In the cell, different copies of the protein may be simultaneously interacting with different partners, may exist in different cellular locations, and may be in different post-translational states, but in the PPIN all these are represented by a single vertex, and all its disparate interactions are represented as undifferentiated outgoing edges from that vertex.

Figure 1 illustrates these shortcomings of the Y2H and TAP-MS methods for detecting PPIs via a simple example; we ignore the effects of other factors such as experimental or biological variability, which in reality would lead to additional false positives (spurious edges) and false negatives (missing edges). Here, we use a simple made-up complex consisting of an A–B–C core, which forms distinct complexes with either

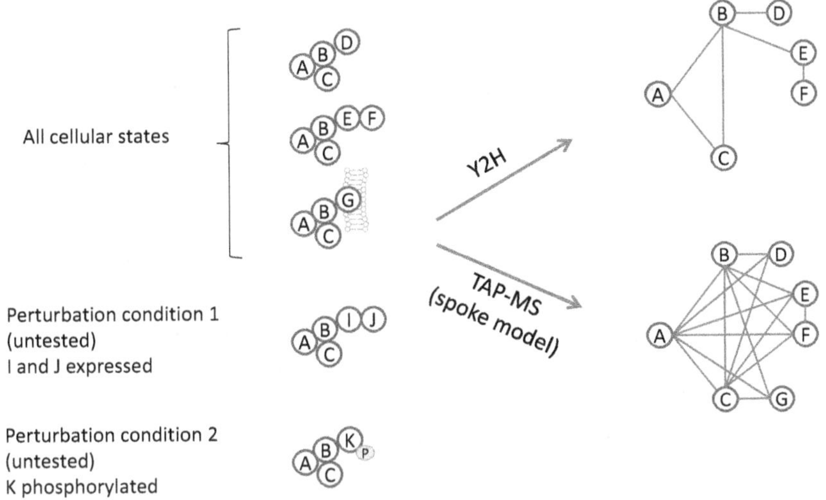

Fig. 1. Detection and representation of dynamically-behaving complexes, in an ideal scenario without spurious or missing interactions. The dynamism of protein complexes is lost after PPI screening and representation in the PPIN. Moreover, this dynamism hinders an accurate screening of PPIs.

protein D, or proteins E–F, or membrane protein G; additionally, it complexes with proteins I–J which are only expressed during perturbation condition 1, and with protein K only after phosphorylation during perturbation condition 2. We assume that all proteins are used as baits in both Y2H and TAP-MS, and in the latter we use the spoke model to obtain individual PPIs. Since the cells interrogated are never in perturbation conditions 1 or 2, proteins I, J, and K are never found to interact with A–B–C. Y2H is unable to detect the interaction with membrane protein G, while the mutually-exclusive interactions with proteins D and E–F are detected and represented as undifferentiated edges. TAP-MS likewise conflates the three distinct complexes as one large, densely-connected graph. While it appears here that the three complexes can be discerned as separate cliques in the graph, in reality the additional spurious and missing edges make this task difficult.

2.5. Augmenting the Static Interactome with Dynamism

Many researchers have recognized that, while the static interactome is a superficial representation of cellular protein interactions, it is still the only proteome-wide and experimentally replicated resource of PPIs that is readily available for computational analysis, and so have attempted to augment it with some degree of dynamism using other information sources.

For example, de Lichtenberg *et al.*[11] integrated yeast PPI data with gene-expression data from various cell-cycle time-points to analyze the dynamism of complex formation during the cell-cycle, and found both constitutively expressed and periodically expressed subunits of most complexes. Likewise, Srihari and Leong[21] also analyzed yeast complexes with cell-cycle expression data, and proposed that constitutively-expressed proteins are likelier to be reused across different complexes.

Other researchers have integrated PPI data with protein-domain information to identify simultaneous or mutually-exclusive interactions. Jung *et al.*[22] decomposed the PPIN into simultaneous protein interaction networks (SPINs), in which all interactions can occur simultaneously, by excluding mutually-exclusive interactions in each SPIN, and then performed complex discovery on each SPIN. Ozawa *et al.*[23] refined predicted complexes by eliminating those that included mutually-exclusive interactions.

A major shortcoming of such analyses is that they are based on the PPIN derived from high-throughput experiments such as Y2H and TAP-MS, so they cannot reveal interactions that are only active in untested conditions.[24] Nevertheless, these approaches show that incorporating this aspect of dynamism in PPIs produces complexes that match known complexes more precisely, and may even elucidate novel functional mechanisms in some complexes. However, the limitations of inferring PPI dynamism indirectly must be noted: for example, gene-expression data does not reflect post-transcriptional activities that further affect complex dynamism, such as protein degradation, transportation, or modification.

3. Three Challenges in Complex Discovery

To discover the set of protein complexes in an organism, researchers have proposed a wide variety of methods to analyze its interactome, derived from high-throughput PPI-screening technologies. A typical strategy is to impute regions of high inter-connectedness in the interactome as putative complexes, since proteins within complexes interact with each other. However, since the basis of this analysis is the static interactome, which as described above lacks crucial information about the dynamism of PPIs, a comprehensive and accurate derivation of complexes becomes problematic.

First, a complex may exist within a highly-connected region of the PPIN, with many extraneous outgoing edges connecting it to other proteins outside the complex. Such a complex is challenging to find, as it is difficult to delimit its boundaries accurately. A particular protein in the complex may have many extraneous PPI edges because it participates in other complexes as well, and the extraneous edges correspond to its interactions with the proteins in these other complexes. These distinct but overlapping (in composition) complexes may exist in different cellular locations, or may form in different cellular states which were detected by the PPI-screening technology, or may even exist in the same location and time as distinct complexes, but this information is not captured in the PPIN. These non-simultaneous interactions corresponding to distinct complexes are active in different copies of the protein, but in the PPIN these multiple copies of the protein are conflated into a single vertex, with all its non-simultaneous interactions corresponding to outgoing edges from that vertex, leading to the many extraneous edges.

The extraneous edges may also correspond to false positives due to a non-physiological environment of the assay, for example through over-expression of bait or prey proteins, or through detected interactions due to post-translational modifications that is different *in vivo*, or through Y2H-detected interactions in the nucleus where the interactors would not localize *in vivo*. Finally, the extraneous edges might simply be an artifact of experimental or other biological variability that is inherent in dealing with biological systems.

Second, a complex may be sparsely-connected in the PPIN, with few PPI edges detected between its proteins. Such a complex does not constitute a dense cluster which can be picked out by clustering algorithms. A complex may be sparse because it is condition-specific: only in certain conditions are its proteins expressed, or modified to enable binding, or co-localized, or the physiochemical environment appropriate for complex formation. If the complex only exists in a condition that was not tested during PPI screening, its proteins' co-complex interactions are not detected. PPIs could also be missing due to technological limitations. Under Y2H, proteins in the complex may not localize in the nucleus or interact in the nucleus where the interaction is assayed — in particular, PPIs in most membrane complexes are not detected. Since Y2H assays interactions in a non-physiological environment, the proteins might not have undergone post-translational modification required for binding, or the environment might be inappropriate for complex formation. Under TAP-MS, weaker interactions may not survive the double-washing step, though they may constitute important interactions within the complex. Finally, as with spurious interactions, missing interactions might also be due to variability in the experimental or biological system.

The third challenge, that of finding small complexes (defined as composed of two or three distinct proteins), is an intrinsic challenge which is exacerbated by the shortcomings of a static interactome. It has been noted that the distribution of complex sizes follows a power law distribution,[25] meaning that a large majority of complexes are small. Thus the discovery of small complexes is an important subtask within complex discovery. An inherent difficulty in this task is that the strategy of searching for dense clusters becomes problematic: fully-dense (i.e. cliques) size-2 and size-3 clusters correspond to edges and triangles, respectively, and only a few among the abundant edges and triangles of the PPIN represent actual small complexes. Furthermore, small complexes are much more sensitive to extraneous or missing edges: For a size-2 complex, a missing co-complex interaction disconnects its two member proteins, while only two extraneous interactions are sufficient to embed it within a larger clique (a triangle). It is apparent that the challenge of small-complex discovery is exacerbated by the two problems of highly-connected regions with many extraneous edges, and sparse regions with many missing edges, in the PPIN. These problems, as described above, owe a great deal to the analysis of a static interactome to derive complexes that are dynamic in nature.

Figures 7 and 8 in the following section illustrate these challenges via two example complexes, and show how they present problems for clustering algorithms.

4. Poor Performance of Current Methods

In this section, we evaluate five clustering algorithms for the prediction of yeast and human complexes. In particular, we highlight three challenges in complex discovery: the prediction of complexes within highly-connected regions of the PPIN, the prediction of sparsely-connected complexes, and the prediction of small complexes.

4.1. *Clustering Algorithms*

In this paper, we evaluate five algorithms, as representatives of different types of clustering algorithms for complex discovery: clique-based, seed-and-grow, simulation, hierarchical, and core-attachment methods. Five additional algorithms — CFinder,[26] IPCA,[27] RNSC,[28] PPSampler,[29] and MCL-CAw[30] — are also evaluated in the Supplementary Materials.

Clustering by Maximal Cliques (CMC[31]) first searches for the set of maximal cliques (cliques that are not contained within a larger clique). Then, for overlapping cliques whose overlap exceeds a threshold, CMC either merges them if they are highly interconnected, or removes the clique with the lower density.

ClusterOne[32] selects vertex seeds based on their degrees, and grows clusters greedily to maximize a cohesiveness function, defined as the ratio of the sum of edge weights within the cluster versus the sum of edge weights within the cluster as well as outgoing edges from the cluster. Furthermore, highly-overlapping clusters are merged.

Markov Clustering (MCL[33]) is based on the principle that a random walker in the PPIN will spend more time traversing a dense region before leaving it. The PPIN is represented as a transition matrix, and the probability of each node visiting every other node at each successive time step is calculated iteratively via matrix multiplication. An inflation step accentuates the differences in probabilities by raising them to a power and then re-normalizing. Regions that are densely connected, with sparse outgoing edges, are found as clusters.

Hierarchical Agglomerative Clustering with Overlap (HACO[34]) first considers all vertices as individual clusters, then iteratively merges pairs of clusters with high connectivity between them. At each merge, the two constituting clusters are remembered; when the merged cluster A is later merged with another cluster B, it also tries to merge the remembered constituting clusters of A with the cluster B, and keeps the (possibly overlapping) resultant clusters if they are highly-connected.

Coach[35] employs a core-attachment model to detect complexes in two stages: core detection and complex formation. In the first stage, neighborhood subgraphs are induced around each vertex and its neighbors, and cores are found as vertices in each neighborhood subgraph that have higher than average local degree, and whose induced subgraph is dense. In the second stage, proteins that are connected to at least some proportion of each core's vertices are recruited as attachments to the core.

Table 1. Five clustering algorithms tested, and their parameters used for discovery of yeast and human complexes.

	Category	**Parameters**
CMC	Clique-based	Yeast: ov = 0.5, mg = 0.5
		Human: ov = 0.5, mg = 0.75
ClusterOne	Seed-and-grow	Yeast and human: default
MCL	Optimization	Yeast: −I 2.5
		Human: −I 4
HACO	Hierarchical	Yeast: −c c 0.75 −g 0.1
		Human: −c c 0.75 −g 0.5
Coach	Core-attachment	Yeast and human: default

Table 1 shows the parameter settings of these five clustering algorithms used for the prediction of yeast and human complexes.

4.2. *Data Sources*

4.2.1. *PPI Data*

A number of repositories for PPI data are available, covering a range of organisms, interactions types (genetic interactions or physical PPIs), interactions sources (such as curated PPIs, experimental PPIs, or predicted PPIs), and experimental detection methods. In our work, we obtain our yeast and human PPIs by taking the union of physical PPIs from three repositories: BioGRID,[36] IntAct,[37] and MINT.[38] In yeast we also incorporate the Consolidated PPI dataset.[17]

We unite these datasets, and score and filter the PPIs, using a simple reliability metric based on the Noisy-Or model to combine experimental evidences (also used by Chua *et al.*[39]). For each experimental detection method e, we estimate its reliability as the fraction of interactions detected where both interacting proteins share at least one high-level cellular-component Gene Ontology term. Then the score of an interaction (a, b) is estimated as:

$$\text{score}(a, b) = 1 - \prod_{i \in E_{a,b}} (1 - \text{rel}_i)^{n_{i,a,b}},$$

where rel_i is the estimated reliability of experimental method i, $E_{a,b}$ is the set of experimental methods that detected interaction (a, b), and $n_{i,a,b}$ is the number of times that experimental method i detected interaction (a, b).

We avoid duplicate counting of evidences across the datasets by using their publication IDs.

Most clustering algorithms perform better when a smaller subset of high-quality PPIs are used. We determined that a score cutoff of 0.8 and 0.7 in yeast and human large complexes respectively, and 0.99 in small complexes, gave decent performance in most clustering algorithms.

4.2.2. *Reference Complexes for Yeast and Human*

To evaluate the performance of complex discovery algorithms, we use reference complexes that have been manually validated via literature curation. For yeast, we use the CYC2008[40] set, which consists of 408 yeast complexes. For human, we use the CORUM[41] set, which consists of 1829 human complexes.

To investigate the performance of the clustering algorithms with respect to the three highlighted challenges, we stratify the reference complexes in terms of their sizes, extraneous edges, and densities. First, to quantify whether a complex is embedded within a highly-connected region of the PPIN, we derive EXT, the number of external proteins that are highly-connected to it, defined as being connected to at least half of the proteins in the complex. Second, to quantify how sparse a complex is, we derive DENS, the density of each complex, defined as the number of PPI edges in the complex divided by the total number of possible edges in the complex. In our analysis, we stratify the complexes into large and small complexes, and further stratify the large complexes into low, medium, and high DENS (corresponding to DENS of $[0, 0.35]$, $(0.35, 0.7]$, and $(0.7, 1]$, respectively), and low and high EXT (corresponding to EXT ≤ 3 and > 3, respectively), to give seven total strata (one for small complexes, and six for large complexes).

Figure 2 illustrates the size distribution of the yeast complexes, and the distributions of EXT, DENS, and our six analysis strata (stratified by EXT and DENS), among the large yeast complexes. Figure 3 shows the corresponding distributions for human complexes. In both yeast and human, the sizes of complexes follow the power-law distribution,[25] which highlights the important subtask of predicting small complexes (of size two and three): among both yeast and human complexes, about 60% are small complexes (259 out of 408 in yeast, 1029 out of 1829 in human).

Among large complexes in both yeast and human, about 40% of complexes have high EXT. We expect the prediction of these complexes

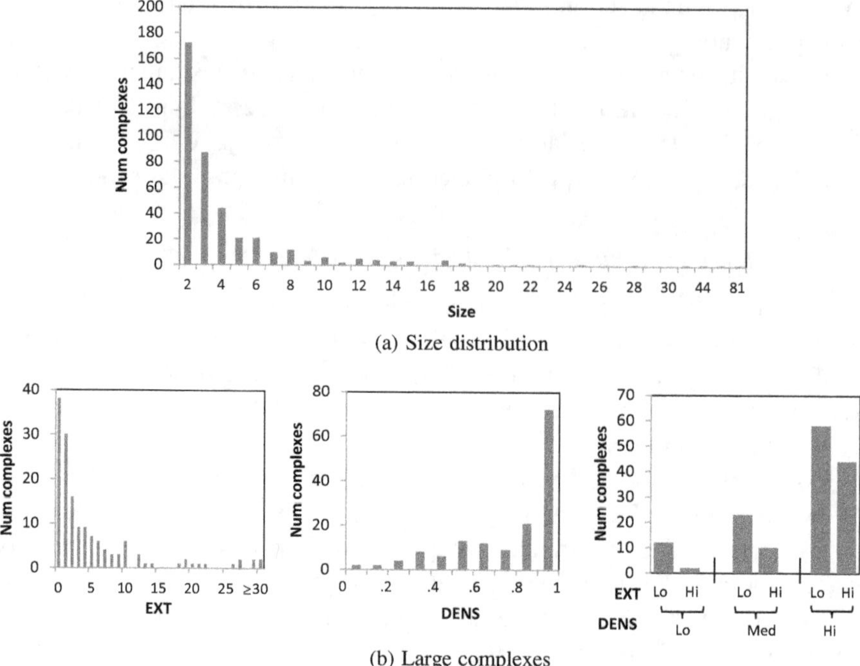

(a) Size distribution

(b) Large complexes

Fig. 2. Statistics of the yeast reference complexes, from the CYC2008 database. (a) The size distribution of the complexes. (b) EXT (number of highly-connected external proteins) and DENS (density) distributions of large complexes.

to be extremely challenging, as it would be difficult to accurately delimit their borders from their highly-connected surroundings (the highly-connected external proteins are likely to be recruited into the predicted complexes). Only 10% of large complexes in yeast have low density. On the other hand, in human about 35% of large complexes are sparsely-connected with low DENS. We expect these sparsely-connected complexes to also be difficult to predict, as they do not form dense clusters that are picked out by most clustering algorithms.

4.3. *Evaluation Methods*

We say that a cluster (i.e. a predicted complex) P matches a known complex C at a given match threshold match_thresh if Jaccard $(P, C) \geq$ match_thresh, where Jaccard (P, C) is the Jaccard similarity

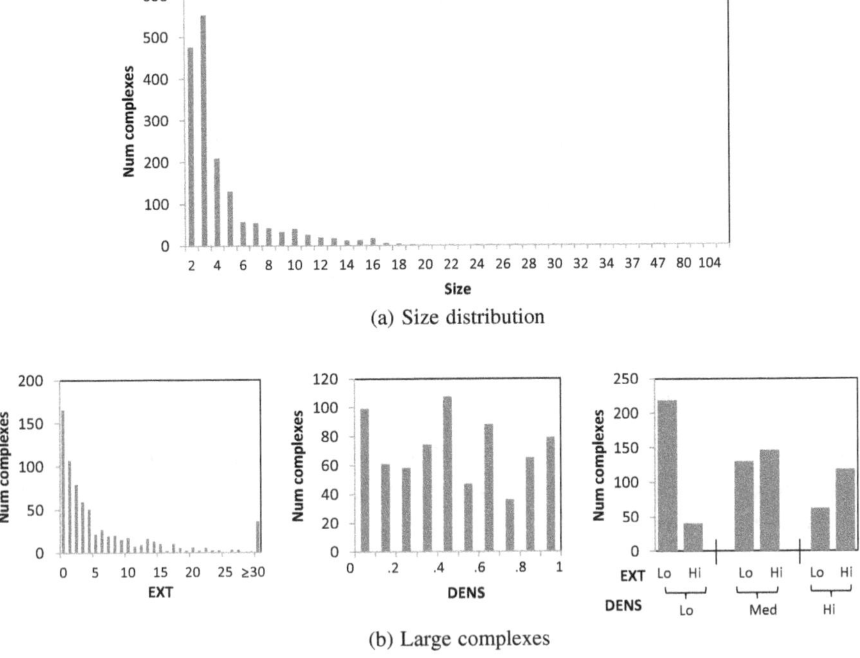

(a) Size distribution

(b) Large complexes

Fig. 3. Statistics of the human reference complexes, from the CORUM database. (a) The size distribution of the complexes. (b) EXT (number of highly-connected external proteins) and DENS (density) distributions of large complexes.

between the proteins contained in P and C:

$$\text{Jaccard}(P, C) = \frac{|V_P \cap V_C|}{|V_P \cup V_C|},$$

where V_X is the set of proteins contained in X. For large complexes, we use a stringent matching criteria of match_thresh $= 0.75$ in matching yeast complexes, and a rougher matching criteria of match_thresh $= 0.5$ in matching human complexes, as the latter task is much more difficult. For small complexes, we use the most stringent criteria of match_thresh $= 1$, as it is easier for a small cluster to match a small complex by chance. Given a set of clusters $\mathbf{P} = \{P_1, P_2, \ldots\}$, and a set of reference complexes $\mathbf{C} = \{C_1, C_2, \ldots\}$, the precision and recall are calculated as:

$$\text{Precision} = \frac{|\{P_i \in \mathbf{P} | \exists\, C_j \in \mathbf{C}, P_i \text{ matches } C_j\}|}{|\mathbf{P}|},$$

$$\text{Recall} = \frac{|\{C_i \in \mathbf{C} | \exists P_j \in \mathbf{P}, P_j \text{ matches } C_i\}|}{|\mathbf{C}|}.$$

The precision–recall graph is another useful measure of performance, as it indicates the quality of predictions at different recall levels. It is obtained by applying varying thresholds on the predicted complexes' weighted densities, and plotting the precision and recall at each threshold. For brevity, we use the area under the precision–recall (AUPR) graph as a summarizing statistic (the actual precision-recall graphs are shown in the Supplementary Materials).

4.4. *Results*

Figures 4(a) and 4(b) show the performance of the clustering algorithms on prediction of large yeast and human complexes, at a finer matching level of match_thresh $= 0.75$ for yeast, and a rougher match_thresh $= 0.5$

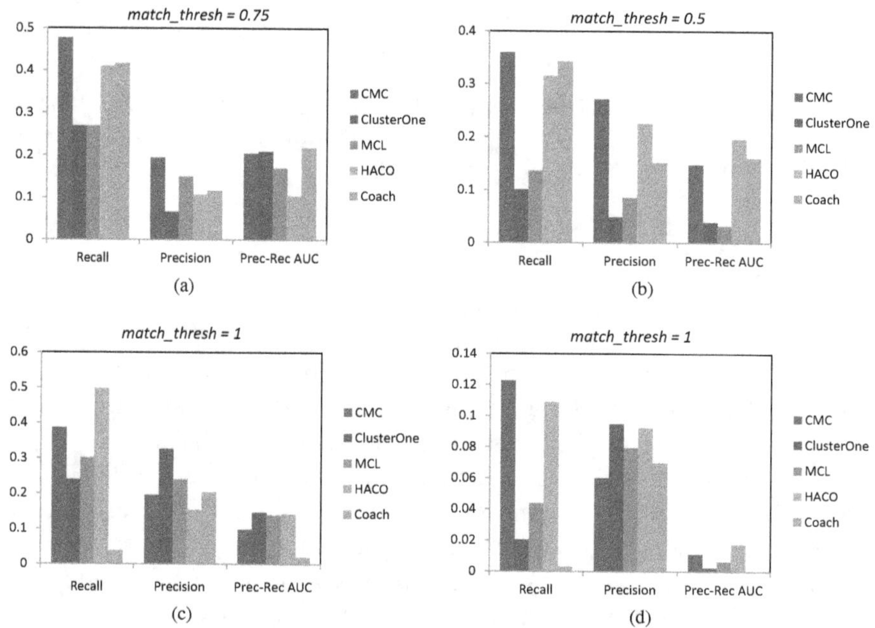

Fig. 4. Performance of the clustering algorithms on prediction of (a) large yeast complexes, (b) large human complexes, (c) small yeast complexes and (d) small human complexes.

for human (as prediction of human complexes is a more difficult task: at match_thresh = 0.75, the highest recall achieved is only about 10%). CMC, HACO, and Coach stand out with the highest recall levels for both yeast and human. In yeast, they achieve recalls over 40%, but CMC attains only 20% in precision, while HACO and Coach do worse with 10% precision. In human, even with a rougher matching criteria, CMC, HACO, and Coach achieve only over 30% recall, with 15% to 25% precision. MCL attains low recall in yeast and human, as it does not generate overlapping clusters. ClusterOne attains the lowest recall and precision, in both yeast and human.

Figures 4(c) and 4(d) show the performance of the clustering algorithms on the prediction of small yeast and human complexes, at a perfect matching requirement of match_thresh = 1.0. Coach predicts fewer than 5% of small yeast complexes, and almost no small human complexes at all, as it is problematic to define tightly connected cores with less-connected attachments when only two or three vertices are available. CMC and HACO again achieve high recall, but at the expense of generating many false positives, as they attain low precision. Again, MCL has low recall as it does not generate overlapping clusters. ClusterOne achieves the highest precision levels for small complexes, but at the expense of the lowest recalls. Note that the performance for small human complexes is dismal: the best clustering algorithm only attains 12% recall.

To investigate which complexes are problematic to predict, we study the performance of the complex discovery algorithms on the complexes stratified in terms of their sizes, extraneous edges, and densities. As described above, the complexes are stratified into small and large complexes, and large complexes are further stratified by density (DENS) and number of highly-connected external proteins (EXT), to give seven groups of complexes (see Figs. 2 and 3 for the distribution of size, DENS, and EXT of yeast and human complexes).

Figure 5(a) shows that yeast complexes with lower density are much harder to predict than those with higher density: no complex with low DENS are predicted at all by any clustering algorithm, while complexes with high DENS are predicted much more frequently. Furthermore, complexes with higher EXT are harder to predict than those with lower EXT: In each density strata, complexes with high EXT have lower recall than those with low EXT. Small complexes are also challenging to predict: Most clustering algorithms do not predict more than 40% of small

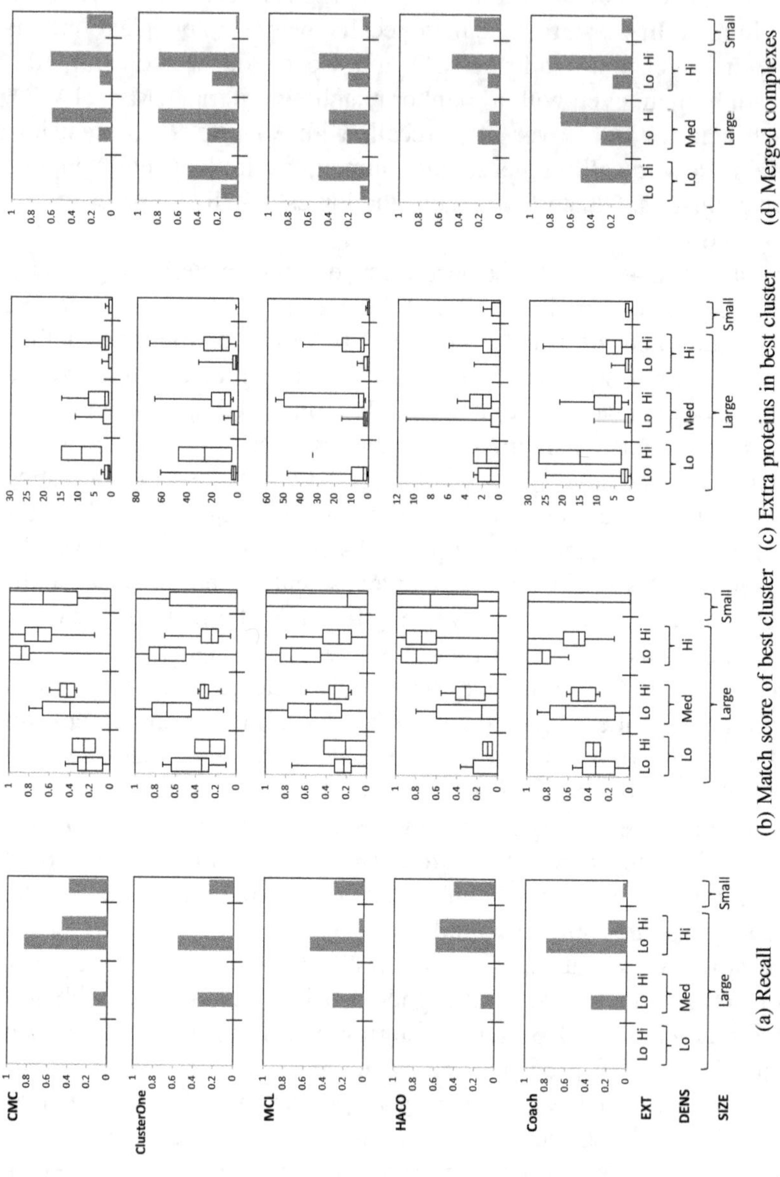

Fig. 5. Performance of complex discovery algorithms on yeast complexes, stratified by size, DENS, and EXT. The x-axis of each chart corresponds to the different stratified groups of complexes, given at the bottom of the figure.

complexes. As expected, the easiest complexes to predict are the large complexes with high DENS and low EXT.

Figure 5(b) shows that complexes with higher density can be predicted with better-matching clusters: Within each EXT strata, the match score increases with density. Furthermore, complexes with lower EXT are predicted with better-matching clusters: among complexes with medium or high DENS, match score is higher among those with low EXT than high EXT.

Figures 5(c) and 5(d) reveal why complexes with higher EXT are difficult to predict. Figure 5(c) shows that clustering algorithms tend to include many extraneous proteins when predicting complexes with higher EXT: Across all DENS strata, complexes with higher EXT have greater number of extra proteins in their best-matched clusters (intuitively, the extraneous proteins are likely to be those highly-connected external proteins). Figure 5(d) shows that clustering algorithms tend to merge together complexes with higher EXT: Across all DENS stratas, complexes with higher EXT tend to be found in clusters merged with other complexes.

Figure 6 shows the corresponding performance of the clustering algorithms on the stratified human complexes. Similar conclusions can be drawn here as from yeast complexes. Small complexes are challenging to predict, with most clustering algorithms predicting less than 10% of them. Complexes with lower density are harder to predict than those with higher density, and are predicted with clusters that match them less well; likewise, complexes with higher EXT are also harder to predict than those with lower EXT, and are also predicted with clusters that match them less well (Figs. 6(a) and 6(b)). However, Fig. 6(b) shows that, within the low-DENS stratum, complexes with high EXT attain slightly higher match scores than those with low EXT, because these low-density complexes with high EXT are likely to slightly overlap with clusters consisting of complex proteins with the external proteins that they are highly-connected to; indeed, in these cases the match scores are mostly under 0.5.

Figure 6(c) shows that, as in yeast, human complexes with high EXT are predicted with clusters that include many more extraneous proteins. Figure 6(d) shows that complexes with higher EXT tend to be merged together in clusters.

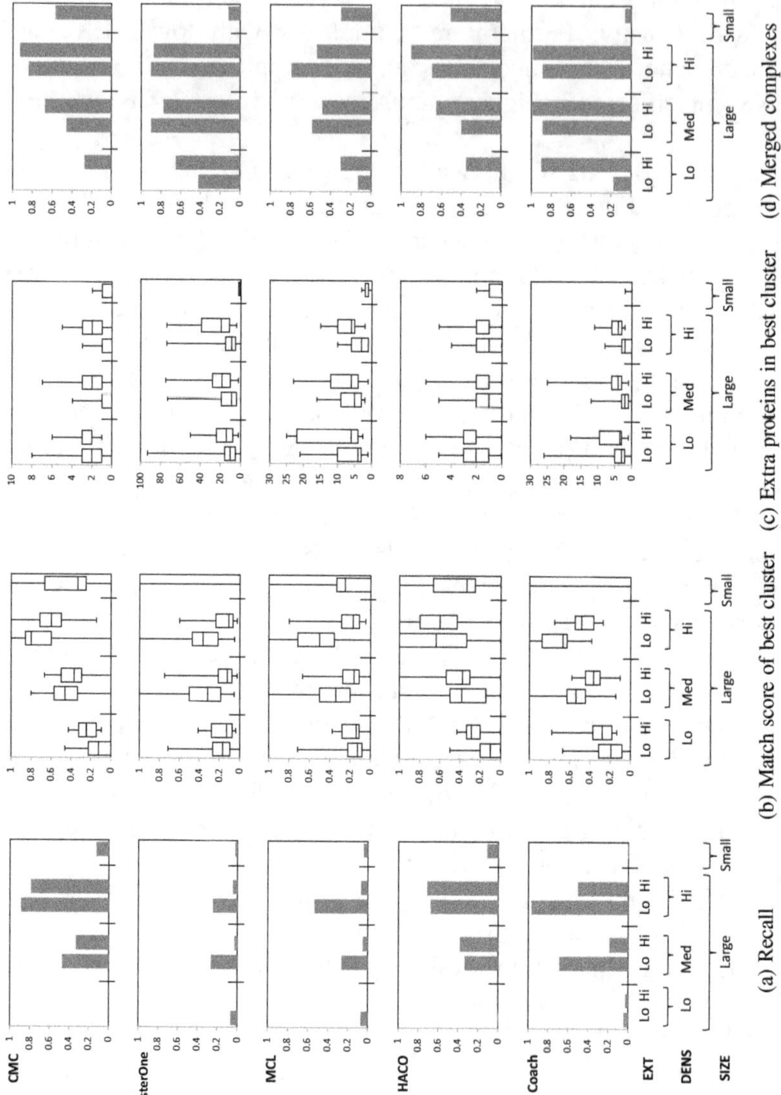

Fig. 6. Performance of complex discovery algorithms on human complexes, stratified by size, DENS, and EXT. The *x*-axis of each chart corresponds to the different stratified groups of complexes, given at the bottom of the figure.

4.5. *Example Complexes*

Here we highlight some example complexes that are known to behave dynamically, and show how their static interactomes exhibit characteristics (such as high EXT and low DENS) which result from their static representation, and which make them difficult to predict.

The Cdc28p yeast protein, as described above, complexes with various cyclin proteins (Cln1p to Cln3p, Clb1p to Clb6p) to regulate the cell-cycle. The proper complexes are formed at each point of the cell-cycle via gene-expression and post-translational controls.[9,10] Figure 7(a) shows the interactome around these proteins and their neighbors, with the nine different complexes formed by Cdc28p circled. Although these interactions occur at different times during the cell-cycle, they are collapsed into the same static interactome, resulting in a highly-connected region around Cdc28p and its cyclin partners: note that the EXT for each of the complexes range from 12 to 13. Furthermore, PPIs are missing between Cdc28p and some of its cyclin partners, giving a density of 0 to these complexes. In fact, these PPIs exist in our source datasets, but with slightly fewer experimental evidences to back them up compared to other Cdc28p PPIs; thus they scored slightly lower in reliability and they were filtered from our PPIN. While it is possible to lower our reliability score cutoff to include these PPIs, this would also include many spurious PPIs and make the discovery of other complexes even more difficult.

Figures 7(b) and 7(c) show the clusters predicted by CMC and MCL, respectively. CMC found four clusters that overlap with four Cdc28p complexes, but with one extraneous protein in each case, while MCL found one large cluster that covered Cdc28p, seven of the nine cyclin proteins, and four extraneous proteins.

The four Replication Factor C (RFC) complexes in yeast are structurally similar complexes involved in DNA metabolism. Each of these complexes consist of a core of four subunits (Rfc2p to Rfc5p) and distinct attachment proteins, and perform different biological functions related to DNA metabolism.[42] The interactome of the RFC complexes and their neighbors are shown in Fig. 8(a), with the four complexes circled. Here again, conflating the four distinct complexes in the static interactome results in many extraneous edges and high connectivity to proteins outside each complex: the EXT for the four complexes range from 4 to 7.

Figures 8(b) and 8(c) show the clusters predicted by CMC and MCL, respectively. CMC predicted one of the RFC complexes perfectly, while

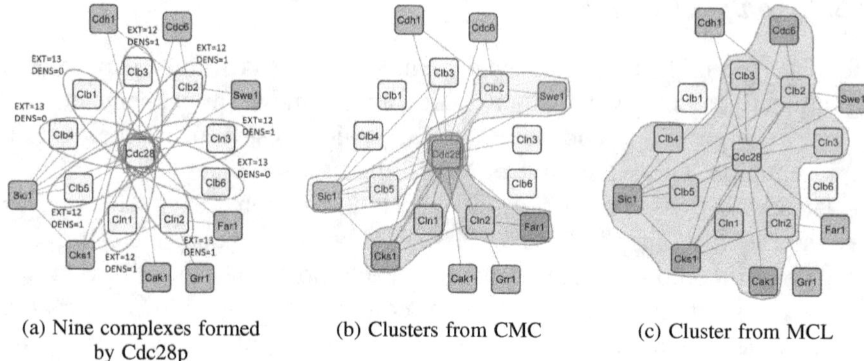

(a) Nine complexes formed by Cdc28p   (b) Clusters from CMC   (c) Cluster from MCL

Fig. 7. (a) Cdc28p is involved in nine distinct complexes, which overlap and have many highly-connected external proteins (EXT). Three of the complexes are disconnected (DENS = 0). (b) CMC includes extraneous proteins in its clusters. (c) MCL merges the complexes.

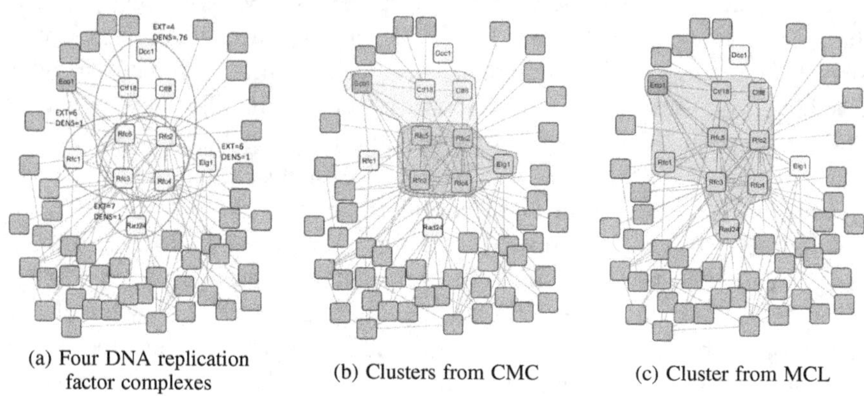

(a) Four DNA replication factor complexes   (b) Clusters from CMC   (c) Cluster from MCL

Fig. 8. (a) A common core is shared among four DNA replication factor complexes, which contributes to a high number of external proteins (EXT) in each complex. (b) CMC finds only one of the four complexes. (c) MCL merges three of the four complexes.

predicting a second cluster that matched another complex less well; MCL predicted a large cluster that overlapped with three of the RFC complexes.

Note that MCL does not allow overlaps in its predicted clusters, so in the above examples it predicts clusters that merge the overlapping and highly-connected complexes together. While CMC allows overlapping clusters, the many extraneous edges and high connectivity to external proteins make it difficult to delimit the overlapping complexes precisely.

5. Discussion and a Call to Arms

Protein interactions behave in a dynamic fashion, with a variety of interaction timings, locations, and affinities. The cellular control of this dynamism gives important functional mechanisms to protein complexes, allowing complexes to assemble at specific times, or to vary in composition to activate or modulate their functions. Interaction detection technologies are limited in their ability to capture such dynamics; furthermore, this dynamism also impedes accurate and comprehensive screening of interactions. Moreover, the representation of interactions in a PPIN does not preserve any information about interaction dynamism, allowing only a static analysis of a dynamic reality.

In Sec. 3, we identified three challenges in complex prediction that result from, and are exacerbated by, the analysis of the static interactome to derive complexes that behave dynamically in nature. First, many proteins participate in multiple complexes, leading to overlapping complexes embedded within highly-connected regions of the PPIN with many extraneous edges connecting them to external proteins. This makes it difficult to accurately delimit the boundaries of such complexes. Second, many condition- and location-specific PPIs are not detected, leading to sparsely-connected complexes that cannot be picked out by clustering algorithms. Third, the majority of complexes are small complexes (made up of two or three proteins), which are extra sensitive to the effects of extraneous edges and missing co-complex edges.

In Sec. 4, we presented results of five clustering algorithms for prediction of large and small complexes in yeast and human, and showed that only complexes with high density and few highly-connected external proteins can be consistently predicted: more than 80% of such large complexes can be predicted in yeast and human (with match_thresh = 0.75 and 0.5, respectively), and more than 60% of such small complexes can be predicted in yeast and human (with match_thresh = 1). Complexes with low density frequently could not be predicted at all, while those with many highly-connected external proteins tended to be predicted in clusters with many extraneous proteins or merged complexes. Furthermore, small complexes with such characteristics are especially challenging to predict, particularly in human for which recall rates are extremely low.

Drawing on our insight into the causes of these challenges, we suggest a few approaches for addressing them.

First, to address the problem of complexes in highly-connected regions with many extraneous edges, Liu *et al.*[43] proposed a technique to decompose the PPIN into spatially- and temporally-coherent subnetworks. First, hub proteins with large numbers of interaction partners are removed before complex discovery, as they tend to correspond to date hubs with non-simultaneous interactions. Next, cellular-location Gene Ontology terms[44] are used to decompose the PPIN into spatially-coherent subnetworks. By splitting dense regions of the PPIN into less-dense but coherent subnetworks, complex-discovery performance is improved, with the biggest improvements among complexes in highly-connected regions. Another reasonable idea to tackle this problem is to make use of information such as protein domains to identify non-simultaneous interactions, which can be used to improve complex discovery. For example, Jung *et al.*[22] decomposed the PPIN into subnetworks of simultaneous interactions, from which temporally-coherent complexes can be extracted; while Ozawa *et al.*[23] refined predicted complexes by eliminating those with non-simultaneous interactions.

Second, to address the problem of sparse complexes, an approach like Supervised Weighting of Composite Networks (SWC[45]) is promising. It integrates PPI data with additional data sources — in particular, functional associations and co-occurrence in literature — using a supervised approach to weight edges with their posterior probability of belonging to a complex. By integrating diverse data sources that may support co-complex relationships between proteins, SWC fills in the missing edges in many sparse complexes, while reducing the amount of spurious non-co-complex edges. Using this approach, improvements are obtained in both precision and recall for yeast and human complex discovery, especially among the sparse complexes.

Third, to address the problem of predicting small complexes, an approach like Size-Specific Supervised weighting (SSS[46]) is promising. It integrates PPI data with two additional data sources, functional associations and co-occurrence in literature, along with their topological features, using a supervised approach to weight edges with their posterior probabilities of belonging to small complexes versus large complexes. SSS then extracts small complexes from the weighted network, and scores them using the probabilistic weights of edges within, as well as surrounding, the complexes. This approach achieves significant improvements in precision and recall in discovering small complexes.

While these approaches perform better than the clustering algorithms surveyed here, there remains much room for improvement, especially for the prediction of human complexes where high levels of accuracy and resolution are still unattainable. We hope that this paper is able to call the bioinformatics community into action to address these highlighted problems, as well as the more general challenge of predicting dynamic protein complexes.

Acknowledgments

This work is supported in part by a Singapore Ministry of Education grant MOE2012-T2-1-061 and a National University of Singapore NGS scholarship.

References

1. Nooren IMA, Thornton JM, Diversity of protein-protein interactions, *EMBO J* **22**(14):3486–3492, 2003.
2. Li X, Wu M, Kwoh CK, Ng SK, Computational approaches for detecting protein complexes from protein interaction networks: A survey, *BMC Genomics* **11**(Suppl 1):S3, 2010.
3. Srihari S, Leong HW, A survey of computational methods for protein complex prediction from protein interaction networks, *J Bioinform Comput Biol* **11**(2):1230002, 2013.
4. Chen B, Fan W, Liu J, Wu FX, Identifying protein complexes and functional modules — From static PPI networks to dynamic PPI networks, *Brief Bioinform* **15**(2):177–194, 2014.
5. Jones S, Thornton JM, Principles of protein-protein interactions, *Proc Natl Acad Sci USA* **93**(1):13–20, 1996.
6. Perkins JR, Diboun I, Dessailly BH, Lees JG, Orengo C, Transient protein-protein interactions: Structural, functional, and network properties, *Structure* **18**(10):1233–1243, 2010.
7. Han JDJ, Bertin N, Hao T, Goldberg DS, Berriz GF, Zhang LV, Dupuy D, Walhout AJM, Cusick ME, Roth FP, Vidal M, Evidence for dynamically organized modularity in the yeast protein-protein interaction network, *Nature* **430**:88–93, 2004.
8. Batada NN, Hurst LD, Tyers M, Evolutionary and physiological importance of hub proteins, *PLoS Comput Biol* **2**(7):e88, 2006.

9. Mendenhall A, Hodge A, Regulation of cdc28 cyclin-dependent protein kinase activity during the cell-cycle of the yeast *Saccharomyces cerevisiae, Microbiol Mol Biol R* **62**(4):1191–1243, 1998.

10. Enserink JM, Kolodner RD, An overview of Cdk1-controlled targets and processes, *Cell Div* **5**(11), 2010.

11. de Lichtenberg U, Jensen LJ, Brunak S, Bork P, Dynamic complex formation during the yeast cell-cycle, *Science* **307**(5710):724–747, 2005.

12. Gavin AC, Aloy P, Grandi P, Krause R, Boesche M, Marzioch M, Rau C, Jensen LJ, Bastuck S, Dumpelfeld B *et al.*, Proteome survey reveals modularity of the yeast cell machinery, *Nature* **440**:631–636, 2006.

13. Fields S, Song O, A novel genetic system to detect protein-protein interactions, *Nature* **340**(6230):245–246, 1989.

14. Brückner A, Polge C, Lentze N, Auerbach D, Schlattner U, Yeast two-hybrid, a powerful tool for systems biology, *Int J Mol Sci* **10**(6):2763–2788, 2009.

15. Rigaut G, Shevchenko A, Rutz B, Wilm M, Mann M, Seraphin B, A generic protein purification method for protein complex characterization and proteome exploration, *Nat Biotechnol* **17**(10):1030–1032, 1999.

16. Gavin AC, Maeda K, Kühner S, Recent advances in charting protein-protein interaction: Mass spectrometry-based approaches, *Curr Opin Biotech* **22**:42–49, 2011.

17. Collins SR, Kemmeren P, Zhao XC, Greenblatt JF, Spencer F, Holstege FCP, Weissman JS, Toward a comprehensive atlas of the physical interactome of *Saccharomyces cerevisiae, Mol Cell Proteomics* **6**(3):439–450, 2007.

18. Krogan NJ, Cagney G, Yu H, Zhong G, Guo X, Ignatchenko A, Li J, Pu S, Datta N, Tikuisis AP *et al.*, Global landscape of protein complexes in the yeast Saccharomyces cerevisiae, *Nature* **440**:637–643, 2006.

19. Bisson N, James DA, Ivosev G, Tate SA, Bonner R, Taylor L, Pawson T, Selected reaction monitoring mass spectrometry reveals the dynamics of signaling through the GRB2 adaptor, *Nat Biotechnol* **29**(7):653–658, 2011.

20. Collins BC, Gillet LC, Rosenberger G, Röst HL, Vichalkovski A, Gstaiger M, Aebersold R, Quantifying protein interaction dynamics by SWATH mass spectrometry: Application to the 14-3-3 system, *Nat Methods* **10**(12):1246–1253, 2013.

21. Srihari S, Leong HW, Temporal dynamics of protein complexes in PPI networks: A case study using yeast cell-cycle dynamics, *BMC Bioinformatics* **13**(Suppl 17):S16, 2005.

22. Jung SH, Hyun B, Jang WH, Hur HY, Han DS, Protein complex prediction based on simultaneous protein interaction network, *Bioinformatics* **26**(3): 385–391, 2010.

23. Ozawa Y, Saito R, Fujimori S, Kashima H, Ishizaka M, Yanagawa H, Miyamoto-Sato E, Tomita M, Protein complex prediction via verifying and reconstructing the topology of domain-domain interactions, *BMC Bioinformatics* **11**:350, 2010.

24. Ideker T, Krogan NJ, Differential network biology, *Mol Syst Biol* **8**:565, 2012.

25. Tatsuke D, Maruyama O, Sampling strategy for protein complex prediction using cluster size frequency, *Gene* **518**(1):152–158, 2012.

26. Adamcsek B, Palla G, Farkas I, Derenyi I, Vicsek T, CFinder: Locating cliques and overlapping modules in biological networks *Bioinformatics* **22**(8):1021–1023, 2006.

27. Li M, Chen J, Wang J, Hu B, Chen G, Modifying the DPClus algorithm for identifying protein complexes based on new topological structures, *BMC Bioinformatics* **9**:398, 2008.

28. Przulj N, Wigle DA, Functional topology in a network of protein interactions, *Bioinformatics* **20**(3):340–348, 2003.

29. Widita CK, Maruyama O, PPSampler2: Predicting protein complexes more accurately and efficiently by sampling, *BMC Syst Biol* **7**(Suppl 6):14, 2013.

30. Srihari S, Ning K, Leong HW, MCL-CAw: A refinement of MCL for detecting yeast complexes from weighted PPI networks by incorporating core-attachment structure, *BMC Bioinformatics* **11**:504, 2010.

31. Liu G, Wong L, Chua HN, Complex discovery from weighted PPI networks, *Bioinformatics* **25**(15):1891–1897, 2009.

32. Nepusz T, Yu H, Paccanaro A, Detecting overlapping protein complexes in protein-protein interaction networks, *Nat Methods* **9**:471–472, 2012.

33. van Dongen S, Graph clustering by flow simulation, PhD Thesis, University of Utrecht, 2000.

34. Wang H, Kakaradov B, Collins SR, Karotki L, Fiedler D, Shales M, Shokat KM, Walther TC, Krogan NJ, Koller D, A complex-based reconstruction of the Saccharomyces cerevisiae interactome, *Mol Cell Proteomics* **8**(6):1361–1381, 2009.

35. Wu M, Li X, Kwoh CK, Ng SK, A core-attachment based method to detect protein complexes in PPI networks, *BMC Bioinformatics* **10**:169, 2009.

36. Chatr-aryamontri A, Breitkreutz BJ, Heinicke S, Boucher L, Winter A, Stark C, Nixon J, Ramage L, Kolas N, ODonnell L, Reguly T, Breitkreutz A, Sellam A, Chen D, Chang C, Rust J, Livstone M, Oughtred R, Dolinski K, Tyers M, The BioGRID interaction database: 2013 update, *Nucleic Acids Res* **41**(Database Issue):D816–D823, 2013.

37. Orchard S, Ammari M, Aranda B, Breuza L, Briganti L, Broackes-Carter F, Campbell NH, Chavali G, Chen C, del Toro N, Duesbury M, Dumousseau M, Galeota E, Hinz U, Iannuccelli M, Jagannathan S, Jimenez R, Khadake J, Lagreid A, Licata L, Lovering RC, Meldal B, Melidoni AN, Milagros M,

Peluso D, Perfetto L, Porras P, Raghunath A, Ricard-Blum S, Roechert B, Stutz A, Tognolli M, van Roey K, Cesareni G, Hermjakob H, The MIntAct project–IntAct as a common curation platform for 11 molecular interaction databases, *Nucleic Acids Res* **42**(Database Issue):D358–D363, 2014.

38. Licata L, Briganti L, Peluso D, Perfetto L, Iannuccelli M, Galeota E, Sacco F, Palma A, Nardozza AP, Santonico E, Castagnoli L, Cesareni G, MINT, the molecular interaction database: 2012 update, *Nucleic Acids Res* **40**(Database Issue):D857–D861, 2012.

39. Chua HN, Sung WK, Wong L, An efficient strategy for extensive integration of diverse biological data for protein function prediction, *Bioinformatics* **23**(24):3364–3373, 2007.

40. Pu S, Wong J, Turner B, Cho E, Wodak SJ, Up-to-date catalogues of yeast protein complexes, *Nucleic Acids Res* **37**(3):825–831, 2009.

41. Ruepp A, Waegele B, Lechner M, Brauner B, Dk I, Fobo G, Frishman G, Montrone C, Mewes H, CORUM: The comprehensive resource of mammalian protein complexes–2009, *Nucleic Acids Res* **38**:D497–D501, 2010.

42. Bylund GO, Majka J, Burgers PMJ, Overproduction and purification of RFC-related clamp loaders and PCNA-related clamps from *Saccharomyces cerevisiae, Metholds Enzymol* **409**:1–11, 2006.

43. Liu G, Yong CH, Chua HN, Wong L, Decomposing PPI networks for complex discovery, *Proteome Sci* **9**(Suppl 1):S15, 2011.

44. Ashburner M, Ball CA, Blake JA, Botstein D, Butler H, Cherry JM, Davis AP, Dolinski K, Dwight SS, Eppig JT *et al.*, Gene ontology: Tool for the unification of biology, *Nat Genet* **25**:25–29, 2000.

45. Yong CH, Liu G, Chua HN, Wong L, Supervised maximum-likelihood weighting of composite protein networks for complex prediction, *BMC Syst Biol* **6**(Suppl 2):S13, 2012.

46. Yong CH, Maruyama O, Wong L, Discovery of small protein complexes from PPI networks with size-specific scoring, *BMC Syst Biol* **8**(Suppl 5):S3, 2014.

Chern Han Yong received his BS and MS degrees in Computer Science from the University of Texas at Austin. He is currently working towards a Ph.D. at the National University of Singapore. His main research interests include computational biology, artificial intelligence, and machine learning.

Limsoon Wong is a Professor in the School of Computing at the National University of Singapore. He currently works mostly on knowledge discovery technologies and their application to biomedicine. He is a Fellow of the ACM, named in 2013 for his contributions to database theory and computational biology. He serves on the editorial boards of several journals, including Journal of Bioinformatics and Computational Biology.

CHAPTER 2

Denoising Protein–Protein Interaction Network Via Variational Graph Auto-Encoder for Protein Complex Detection[a]

Heng Yao[*], Jihong Guan[†] and Tianying Liu[‡]

*Department of Computer Science and Technology
Tongji University, 4800 Cao'an Road
Shanghai 201804, P. R. China*

*Key Laboratory of Embedded System
and Service Computing (Tongji University)
Ministry of Education, Shanghai, P. R. China*

*Shanghai Electronic Transactions and Information Service
Collaborative Innovation Center
Shanghai, P. R. China*
[*]*yaoheng@tongji.edu.cn*
[†]*jhguan@tongji.edu.cn*
[‡]*liutianying@tongji.edu.cn*

Identifying protein complexes is an important issue in computational biology, as it benefits the understanding of cellular functions and the design of drugs. In the past decades, many computational methods have been proposed by mining dense subgraphs in Protein–Protein Interaction Networks (PINs). However, the high rate of false positive/negative interactions in PINs prevents accurately detecting complexes directly from the raw PINs. In this paper, we propose a denoising approach for protein complex detection by using variational graph auto-encoder. First, we embed a PIN to vector space by a stacked graph convolutional network

---

[a]This article was previously published in *Journal of Bioinformatics and Computational Biology*. Vol: 18, No. 3 (2020), 2040010 (28 pages).

[†]Corresponding author.

(GCN), then decide which interactions in the PIN are credible. If the probability of an interaction being credible is less than a threshold, we delete the interaction. In such a way, we reconstruct a reliable PIN. Following that, we detect protein complexes in the reconstructed PIN by using several typical detection methods, including CPM, Coach, DPClus, GraphEntropy, IPCA and MCODE, and compare the results with those obtained directly from the original PIN. We conduct the empirical evaluation on four yeast PPI datasets (Gavin, Krogan, DIP and Wiphi) and two human PPI datasets (Reactome and Reactomekb), against two yeast complex benchmarks (CYC2008 and MIPS) and three human complex benchmarks (REACT, REACT_uniprotkb and CORE_COMPLEX_human), respectively. Experimental results show that with the reconstructed PINs obtained by our denoising approach, complex detection performance can get obviously boosted, in most cases by over 5%, sometimes even by 200%. Furthermore, we compare our approach with two existing denoising methods (RWS and RedNemo) while varying different matching rates on separate complex distributions. Our results show that in most cases (over 2/3), the proposed approach outperforms the existing methods.

Keywords: Protein–Protein Interaction Network (PIN); protein complex; graph embedding; variational auto-encoder.

1. Background

A protein complex is a group of interacting proteins (usually made of polypeptide chains). Protein complexes play essential roles in living organisms. Although there are some experimental techniques such as Tandem Affinity Purification with Mass Spectrometry (TAP-MS) that can identify protein complexes directly, such experimental methods are expensive and inefficient. What is more, owing to the technical limitation, it is difficult to detect complexes consisting of weak interactions. In the past decades, with the development of high-throughput approaches [e.g. yeast two-hybrid screen (Y2H) and tandem affinity purification (TAP)], people can obtain Protein-Protein Interaction (PPI) data in a very large scale. Therefore, many computational methods were proposed to identify protein complexes from PPI data. In these methods, PPI data are represented as a network, called Protein–Protein Interaction Network (PIN), where nodes correspond to proteins and edges correspond to interactions between proteins. Then, protein complex detection turns to mine densely-connected subnetworks in PINs.

Early methods assume that each protein participates in one and only one complex, so traditional graph clustering approaches can be applied to complex detection.[1,2] However, such an assumption is not realistic. Thus, overlapping clustering strategies were employed in later studies, where the effects of neighboring proteins[3–5] and core-attachment model[6,7] were considered. Still, most of these methods cannot achieve satisfactory detection results due to the high rate of false positive/negative interactions in PINs. In recent years, some researchers tried to mine high-order topological information of PINs to evaluate protein-protein relationship.[8–11] Others exploited additional information [e.g. gene expression, gene ontology (GO), protein domains and functions] to enhance protein interactions.[12–14] Besides these methods above, there are also a series of complex detection methods[15–19] that adopt a "from function to interaction" idea.

For protein complexes detection, apart from enhancing PINs by topology or additional features of PINs, some works filtered out false connections directly.[8,20,21] In these works, PPIs were evaluated by adopting explicit features but ignoring the implicit features of PINs. Recently, graph embedding, aiming to represent nodes in graphs as vectors, has gained much attention in graph mining such as graph clustering.[22] There are several typical graph embedding methods, including DeepWalk,[23] Node2Vec,[24] LINE[25] and SDNE.[26] As these self-learning methods could discover potential features of graphs, they become an alternative approach to mining underlying structures of graphs, and have the ability to promote the performance of complex detection.

In this paper, we propose a novel method for detecting protein complexes by denoising PPI networks with a graph convolutional network (GCN). Concretely, we first embed a PIN to vector space via variational graph auto-encoder (VGAE). Then for each interaction in the original PIN, we calculate the probability of being credible. Next, we reconstruct the PIN with interactions with the probability larger than a threshold. Finally, we detect protein complexes on the reconstructed PIN by using six typical methods including CPM, Coach, DPClus, GraphEntropy, IPCA and MCODE. On yeast, we conduct experiments on four PPI datasets (Gavin,[27] DIP,[28] Krogan[29] and Wiphi[30]) against two benchmark complex datasets (CYC2008[31] and MIPS[32]); on human, we evaluate the performance of PPI datasets (Reactome[33] and a subset Reactomekb) against complex datasets (REACT,[33] REACT_uniprotkb and CORE_COMPLEX_human[34]). Experimental results show that performance on the reconstructed PINs is obviously superior to that on the original PINs. Moreover, we compare our denoising method with two existing ones, and validate the superiority of the proposed method. In Sec. 4, we refer to Ref. 35 to illustrate our results with different complex groups according to the size of a complex, the number of highly linked external proteins and the density of a complex.

The rest of this paper is organized as follows: Section 2 surveys related works. Section 3 presents the method. Section 4 evaluates the proposed method, and finally Sec. 5 concludes the paper.

2. Related Works

In this section, we briefly review the related works of this paper from three aspects: (1) denoising PINs; (2) graph embedding; and (3) variational graph auto-encoder.

2.1. *Denoising PINs*

For complexes detection, most existing methods strive to find densely-connected subgraphs directly from PINs, such as Clique,[5] CFinder,[3] and those based on the core-attachment model (e.g. Core[6] and Coach[7]). As we can see, the quality of PINs profoundly impacts the performance of these methods. Thus, some studies exploit topological or functional information to exclude the edges of low confidence.[8,20,21] In PCP[8] assuming that a protein and its second-order neighbors always share common biological functions, the authors define a kind of topological weight (*FS-Weight*) to estimate the strength of interactions, and the interactions of low weight are removed. In SLPC,[21] the interactions whose GO similarity is less than 0.9 are deemed false positive and deleted from the original PINs. In Ref. 20, with a view of both topology and functions, a hybrid similarity is defined to filter out interactions of low reliability.

There are also works that predict new PPIs based on the topology of a PIN. Kuchaiev *et al.*[36] discovered potential interactions by a geometric denoising approach. First, embedding a PIN by multi-dimensional scaling (MDS), then computing the Euclidean distance of embedding. If the distance of a certain pair of proteins is less than a predefined threshold, an interaction is added between this pair of proteins. There are also works that remove spurious interactions and predict new interactions. RWS[37] calculates similarity by a random walk with resistance; RedNemo[38] repeats diffusion with neighborhood modifications iteratively. In this study, we aim to leverage the variational deep neural network model to filter out interactions of low reliability in order to boost the protein complex detection.

2.2. *Graph Embedding*

As a novel graph representation method, graph embedding attempts to embed a graph to vector space, where each vector represents a node (or a subgraph in some definitions). Several works[39,40] showed that graph embedding can mine the high-order structures (e.g. community and degree distribution) of PINs. In this paper, graph embedding is utilized to evaluate the probability that two proteins credibly interacted. Up to now, several typical graph embedding algorithms have been proposed. DeepWalk[23] obtains sequences from a truncated random walk, and captures the features by a language model. Node2Vec[24] extends the DeepWalk method by adding a parameter to balance BFS-like walking and DFS-like walking. Tang *et al.*[25] tried to embed networks using a probabilistic scheme while Wang *et al.*[26] manipulated the neural network. All these methods can be regarded as an encoder–decoder architecture: the original PIN is fed into any auto-encoder, the code of certain dimension is the latent representation. Finally, the code is decoded to reconstruct the original network. In the process of encoding and decoding, the model usually maintains a predefined similarity.

2.3. *Variational Graph Auto-Encoder*

Variational graph auto-encoder,[41] proposed by Kipf and Welling, is a variant of auto-encoder built for graph data rather than image data or text data. Inspired by spectral graph convolutions[42] with Eigenvalue Decomposition, the graph convolutional network (GCN) does the first-order approximation by a truncated expansion in terms of Chebyshev polynomials. The time complexity is reduced from square to linear. In VGAE, the GCN layer is used as the encoder, an inner product of pairwise vectors is used as the decoder. Different from graph auto-encoder, VGAE further introduces a stochastic latent variable z to exploit the unseen features in networks ostensibly. Lately, a few works extended VGAE by marginalizing the corrupted features[43] or embellishing it in an adversarially regularized model.[44]

3. Method

In this section, we first give an overview of our method, then present the technical details.

3.1. *Overview*

Figure 1 shows the workflow of our method, which consists of four major steps:

- **Step-1:** *Embedding a PIN to vector space.* Given a PIN (or a PPI dataset), we do graph embedding by VGAE: as the input of a three-layer

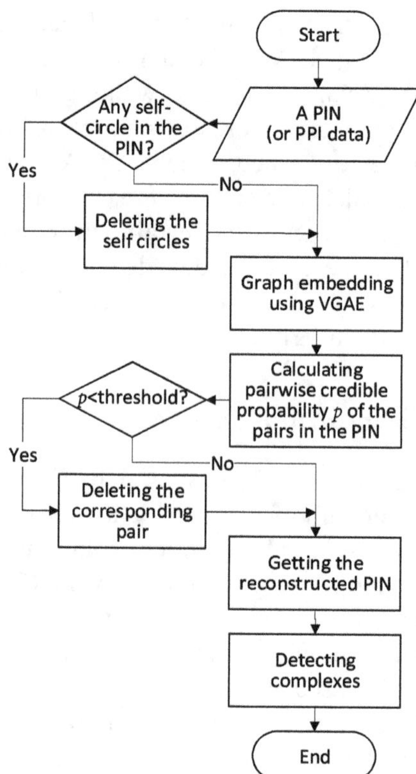

Fig. 1. The workflow of our method. It consists of four major steps: (1) Embedding a PIN to vector space; (2) calculating pairwise credible probability; (3) reconstructing the PIN; and (4) detecting protein complexes from the reconstructed PIN.

graph convolutional network, the PIN is embedded to vectors with each vector corresponding to a node of the PIN.

- **Step-2:** *Calculating pairwise credible probability.* After obtaining the proteins' feature vectors, we calculate the credible probability of each interacting protein pair. A larger probability means a more credible interaction. Thus, we get a table of credible probabilities of all protein pairs.
- **Step-3:** *Reconstructing the PIN.* We set a threshold to discriminate positive/false interactions in the original PIN. Interactions with probability no less than the threshold are taken as positive interactions or credible interactions; otherwise, they are taken as false interactions, which are removed from the original PIN. All credible interactions constitute a new PIN, which will be used for complex detection later.
- **Step-4:** *Detecting protein complexes from the reconstructed PIN.* Over the reconstructed PIN, we detect protein complexes using six existing methods, including MCODE, DPClus, CPM, IPCA, Coach and GraphEntropy.

3.2. *Embedding a PIN to Vector Space*

We utilize VGAE built with a three-layer GCN to embed a PIN. Before giving the details of VGAE, we briefly introduce GCN. Actually, GCN is a simplified class of graph convolutional neural networks based on spectral graph theory.[42,45]

The convolution operation in spectral graph convolutional neural networks is defined as the product of g_θ and x, where x is the input data and g_θ is a signal filter, $\star$ represents the convolution operation. Formally,

$$g_\theta \star x = U g_\theta U^T x, \tag{1}$$

where $U^T x$ is the graph Fourier transform of x, U comes from the normalized graph Laplacian as follows:

$$L = I - D^{-1/2} A D^{-1/2} = U \Lambda U^T, \tag{2}$$

where Λ is the eigenvalues diagonal matrix. The filter g_θ can be understood as a function of the eigenvalues of L, i.e. $g_\theta = g_\theta(\Lambda)$. Considering the high time complexity of eigenvector computation, Defferrard *et al.*[45] applied Chebyshev polynomials $T_k(x)$ up to kth order to approximate $g_\theta(\Lambda)$:

$$g_\theta(\Lambda) \approx \sum_{k=0}^{K} \theta'_k T_k(\hat{\Lambda}), \tag{3}$$

with $\hat{\Lambda} = \frac{2}{\lambda_{\max}}\Lambda - I_N$. $\lambda_{\max}$ is the largest eigenvalue of L, θ' is a vector of Chebyshev coefficients, and $T_k(\cdot)$ denotes the Chebyshev polynomials.

Now going back to Eq. (1), and replacing g_θ by $g_\theta(\Lambda)$, we get

$$g_\theta \star x \approx \sum_{k=0}^{K} \theta'_k T_k(\hat{L})x, \tag{4}$$

with $\hat{L} = \frac{2}{\lambda_{\max}}L - I_N$. Here, $U(\Sigma_{k=0}^{K}\theta'_k T_k(\hat{\Lambda}))U^T = \Sigma_{k=0}^{K}\theta'_k T_k(U\hat{\Lambda}U^T)$.

By limiting $K = 1$ and $\lambda_{\max} = 2$, Kipf and Welling[46] further approximated this equation as follows:

$$g_\theta \star x = \theta(I + D^{-\frac{1}{2}}AD^{-\frac{1}{2}})x \approx \theta(\hat{D}^{-\frac{1}{2}}\hat{A}\hat{D}^{-\frac{1}{2}})x, \tag{5}$$

where $\hat{A}$ is the sum of identity matrix I and adjacency matrix A, $\hat{D}$ is the degree matrix of $\hat{A}$. Let $\tilde{A} = \hat{D}^{-\frac{1}{2}}\hat{A}\hat{D}^{-\frac{1}{2}}$, we can formally generalize the GCN model as

$$\mathrm{GCN}(X, A) : H^{(l+1)} = \mathrm{ReLU}(\tilde{A}H^{(l)}W^{(l)}), \tag{6}$$

where X is the input feature matrix, $H^{(l)}$ ($H^{(0)} = X$) is the active matrix of the lth layer, W is the trainable weight matrix of the lth layer and $\mathrm{ReLU} = \max(0, \cdot)$ is the active function of GCN.

VGAE is parameterized by a multi-layer GCN, whose inference model is defined as

$$q(Z|X, A) = \prod_{i=1}^{N} q(z_i|X, A), \tag{7}$$

where X conforms to a Gaussian distribution: $q(z_i|X, A) = \mathcal{N}(z_i|\mu_i, \mathrm{diag}(\sigma_i^2))$. Z is a latent variable, μ_i is the mean vector and σ_i^2 is the variance vector. Both μ_i and σ_i^2 are derived by GCN, and in practice, we choose to optimize the logarithmic variance $\log \sigma_i^2$.

The loss function of VGAE is as follows:

$$\mathcal{L} = \mathbb{E}_{q(Z|X,A)}[\log p(A|Z) - \mathrm{KL}[q(Z|X, A)\|p(Z)]], \tag{8}$$

where we use a Gaussian prior $p(Z) = \Pi_i \mathcal{N}(z_i|0, I)$, and $\mathrm{KL}[q\|p]$ is the Kullback–Leibler divergence between distribution q and distribution p.

Note that in experiments, we do not add any extra features, thus replace X by the identity matrix.

To obtain robust embeddings of nodes in PINs, we consider the idea of denoising auto-encoder,[47] the edges in the original PIN are removed randomly by a certain rate(de_{rate}) while training the VGAE model. As listed in Table 3, the denoising rate is set to 0.05.

3.3. *Calculating Pairwise Credible Probability*

The generative model of VGAE is defined as the probability to reconstruct A, given the latent variable Z:

$$p(A|Z) = \prod_{i=1}^{N} \prod_{j=1}^{N} p(A_{ij}|z_i, z_j). \tag{9}$$

After getting embeddings z_i and z_j, which can be regarded as self-learned features, we calculate pairwise credible probability of node i and node j as follows:

$$p(A_{ij} = 1|z_i, z_j) = \text{sigmoid}(z_i^T z_j). \tag{10}$$

3.4. *Reconstructing the PIN*

We set a threshold to discriminate whether the interaction between node i and node j is credible ($A_{ij} = 1$) or incredible ($A_{ij} = 0$). If the credible probability of pair (z_i, z_j) is equal to or greater than a threshold, this connection is preserved; otherwise, we delete it. With all credible interactions, we get a new PIN, which is the reconstructed PIN. As all incredible interactions (or noise) are removed, the reconstructed PIN is more reliable than the original one.

3.5. *Detecting Protein Complex on the Reconstructed PIN*

With the reconstructed PIN, six typical existing methods are employed to detect protein complexes, including MCODE, DPClus, CPM, IPCA, Coach and GraphEntropy. Here, we give a concise introduction to all these six methods.

- MCODE[4]: As a well-known method of complex detection, MCODE first weights a vertex by its neighborhood (k-core, defined as the minimum degree of an adjacent subgraph), then implements a seed-growth strategy to predict protein complexes: seeds are selected by the weights of vertexes, and are grown by a predefined vertex weight percentage (denoted as vwp). After detection, post-processing includes removing complexes whose size is less than 2, and adding densely-connected neighbors of candidate complexes.

- DPClus[48]: It converts an unweighted PIN to a weighted one first. The number of common neighbors is used as the edge weight of two proteins, and the weighted degree is defined as the node weight. As a kind of seed-growth methods, instead of stopping growth by the node weight like MCODE, DPClus stops by two thresholds: cluster density (d_{in}), and cluster property (cp_{in}). In DPClus, once a candidate complex is discovered, all weights are updated by the remaining network. To consider overlapping of proteins, DPClus also expands the candidate complexes by possible neighbors.

- CPM[49]: Up to now, several clique percolation algorithms have been proposed for detecting dense communities in social networks and biological networks, one is CPM. The first step of CPM is to find all k-order cliques, which are defined as subgraphs fully connected to each other, and consisting of k nodes. Then, the cliques are percolated together if they can be *reachable* until none of the community is *reachable*. Here, *reachable* means that in these cliques, at least $k - 1$ nodes are shared.

- IPCA[50]: As a variant of DPClus, IPCA uses a similar strategy to compute edge/node weights at the beginning, but does not update the weights iteratively. In expanding the candidate complexes, IPCA selects neighbors according to two conditions: (1) the ratio of *#links* to *size* is not less than the minimum neighbor percentage (T_{in}). Here, *#links* denotes the number of links between a neighbor and the candidate complex, and *size* denotes the number of proteins in a candidate complex. (2) The path length between a neighbor and any member of the candidate complex is not larger than a threshold: maximum shortest path length (d).

- Coach[7]: Based on the core-attachment model, Coach consists of two major steps: (1) finding cores of the potential complexes and (2) adding strongly associated neighbors. In the first step, two parameters are specified: minimum core density (d) and maximum core affinity (t). These two parameters are used to mine the preliminary cores. In the second step, one parameter, minimum neighbor closeness (c), is set to attach a noncore neighbor to cores.

- GraphEntropy[51]: This is an information theory-based method for protein complex detection. It formalizes the measure of graph entropy as the sum of all nodes' entropies in a PIN. It first randomly selects a seed set and forms a seed cluster (a node and its neighbors). Then, it deletes nodes within the seed cluster and adds nodes outside the seed cluster so that the graph entropy decreases, until the seed set is completely traversed.

4. Performance Evaluation

We conduct extensive experiments to evaluate the proposed method. Here, we first present the PPI datasets, complex benchmarks and performance metrics. Then, we describe the parameter settings of VGAE and all detecting methods. Next, we check the effect of the threshold of credible interactions. After that, we compare the complex detection performance on the reconstructed PINs and the original PINs. Finally, we compare our method with two existing PPI denoising algorithms while varying the matching rate.

4.1. *Datasets*

In our experiments, we use four yeast PPI datasets (DIP,[28] Gavin,[27] Krogan[29] and Wiphi[30]) and two human PPI datasets (Reactomekb and Reactome, extracted from the Reactome database[b]). For yeast PPI datasets, we benchmark them by two complex datasets CYC2008.[31] and MIPS[32] In addition, the Reactome dataset is benchmarked by REACT while Reactomekb is benchmarked by REACT_uniprotkb and CORE_COMPLEX_human.[c] Tables 1 and 2 give the basic statistics of the PPI datasets and the benchmark datasets, respectively.

To better investigate the performance of the clustering algorithms on the reconstructed PINs, we also adopt the evaluation metrics proposed by Yong and Wang[35]: SIZE, DENS and EXT. The SIZE is defined as the number of proteins within a protein complex. If a complex contains two or three proteins ($\text{SIZE} \in [2, 3]$), it is classified into small complexes group;

---

[b] https://reactome.org/download-data.

[c] https://mips.helmholtz-muenchen.de/corum/.

Table 1. The statistics of PPI datasets.

PPI datasets	No. of proteins	No. of interactions	Avg. degree
DIP	4928	17,201	7.0
Gavin	1855	7669	8.3
Krogan	2674	7075	5.3
Wiphi	5955	50,000	16.8
Reactomekb	5397	21,021	7.8
Reactome	8141	32,203	7.9

Table 2. The statistics of complex benchmark datasets.

Benchmark datasets	No. of proteins	No. of complexes
CYC2008	1627	349
MIPS	1237	313
REACT_uniprotkb	7823	8499
REACT	10,368	12,429
CORE_COMPLEX_human	3449	2417

otherwise, it is a large complex. The DENS is defined as the number of existing edges in the complex divided by the total number of possible edges in the complex. The EXT is defined as the number of highly connected external proteins out of the complex, "highly" means an external protein links at least half of the proteins in the complex. In general, the large complexes fall into three groups of DENS: low (DENS $\in [0, 0.35]$), medium (DENS $\in (0.35, 0.70]$) and high (DENS $\in (0.70, 1.0]$). Likewise, the large complexes are also ordered by low external ratio and high external ratio while their EXT ≤ 3 and EXT > 3, respectively.

In the following, we present the complex distributions according to their SIZE, DENS and EXT. For yeast, the reference complexes are verified according to CYC2008 with interactions extracted from Wiphi (see Fig. 2). For human, the reference complexes are visualized according to Reactome with interactions extracted from REACT (see Fig. 3).

From Figs. 2 and 3, we find that the SIZE obeys the power-law distribution, meaning that small ones are the majority of discovered complexes. Although many complexes are aranged in low EXT (relatively, easy to be detected), a large amount of complexes still have a high EXT. DENS distribution is different apparently, the proportion of high DENS complexes in REACT is larger than that in CYC2008, which denotes the

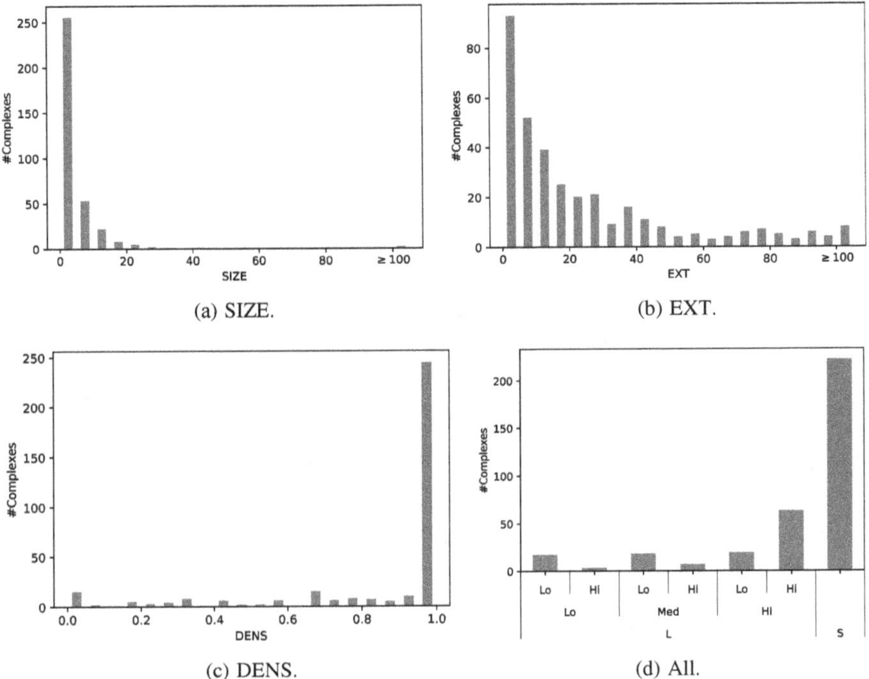

(a) SIZE.

(b) EXT.

(c) DENS.

(d) All.

Fig. 2. Complexes distributions of CYC2008 with interactions extracted by Wiphi. In panel (d), the labels of x-axis, from top, center to bottom, are EXT, DENS and SIZE, respectively.

extremely challenging difficulty of detecting real complexes on human PPI datasets.

4.2. *Performance Evaluation Metrics*

The match rate between a predicted complex and a real one is defined as

$$\text{Rate}_{\text{match}} = \frac{|\text{pc} \cap \text{rc}|^2}{|\text{pc}| \times |\text{rc}|}, \tag{11}$$

where pc and rc are the sets of proteins of the predicted complex and the benchmark complex. As in the existing works, if $\text{Rate}_{\text{match}} \geq 0.2$, we regard that the predicted complex is matched to the benchmark complex (to strengthen the confidence, we also describe the results in higher $\text{Rate}_{\text{match}}$, see Sec. 4.6.2). We adopt three performance metrics, recall,

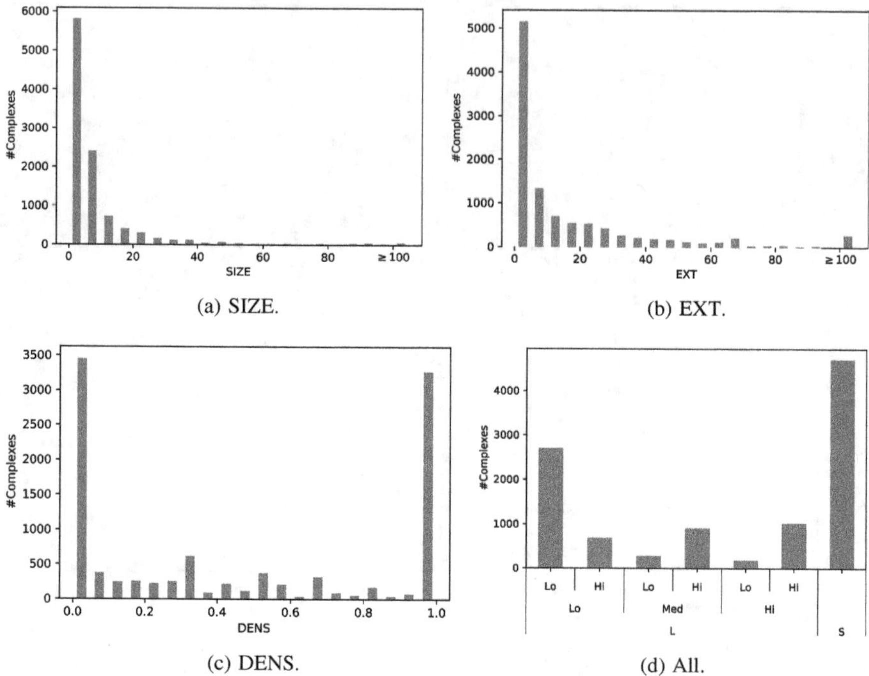

Fig. 3. Complexes distributions of REACT with interactions extracted by Reactome. In panel (d), the labels of x-axis, from top, center to bottom, are EXT, DENS and SIZE, respectively.

precision and $F1$-measure (or $F1$ in short), which are evaluated as follows:

$$\text{recall} = \frac{n_{\text{rc}}}{N_{\text{rc}}}, \tag{12}$$

$$\text{precision} = \frac{n_{\text{pc}}}{N_{\text{pc}}}, \tag{13}$$

$$F1 = \frac{2 \times \text{precision} \times \text{recall}}{\text{precision} + \text{recall}}. \tag{14}$$

Above, N_{rc} is the number of real complexes (benchmark complexes), and n_{rc} is the number of real complexes that match at least one predicted complex. N_{pc} is the number of predicted complexes, and n_{pc} is the number of predicted complexes that match at least on real complex. $F1$ is the harmonic mean of recall and precision.

4.3. *Parameter Settings in VGAE and Detecting Methods*

The parameter settings of VGAE are listed in Table 3. The dimension of embeddings is set to 32, which is denoted by l_2. The learning rate is set to 0.01 to speed up the training process. To acquire robust embeddings of PINs, we set $de_{rate} = 0.05$. To prevent overfitting, we drop out the neurons in the rate of 0.1.

In Table 4, we present the parameter settings of the six complex detection methods. Following Ref. 49, we keep the default parameters in the original articles of these methods. In the following experiments, all the parameters' values are kept the same.

Table 3. The parameter settings of VGAE.

Method	Parameters	Settings
VGAE	Learning rate (η)	$\eta = 0.01$
	Training epochs (E)	$E = 300$
	Dimension of hidden layer 1 (l_1)	$l_1 = 256$
	Dimension of hidden layer 2 (l_2)	$l_2 = 32$
	Dropout rate ($drop_{rate}$)	$drop_{rate} = 0.1$
	Denoising rate (de_{rate})	$de_{rate} = 0.05$

Table 4. The parameter settings of complex detection methods.

Methods	Parameters	Settings
CPM	Clique size (k)	$k = 3$
Coach	Minimum core density (d)	$d = 0.7$
	Maximum core affinity (t)	$t = 0.225$
	Minimum neighbor closeness (c)	$c = 0.5$
DPClus	Cluster density (d_{in})	$d_{in} = 0.9$
	Cluster property (cp_{in})	$cp_{in} = 0.5$
GraphEntropy	N/A	N/A
IPCA	Maximum shortest path length (d)	$d = 2$
	Minimum neighbor percentage (T_{in})	$T_{in} = 0.5$
MCODE	Vertex weight percentage (vwp)	vwp $= 0.2$

4.4. *Effect of Credible Probability Threshold*

To check the effect of the credible probability threshold, we vary the threshold from 0 to 0.9, and present the statistics of the resulting reconstructed yeast PINs in Table 5. Figure 4 illustrates the ratios of nodes/ edges in the reconstructed PINs over those of the corresponding original PINs. From Table 5 and Figure 4, we can see that when the threshold falls

Table 5. The statistics of reconstructed PINs.

Threshold	DIP		Gavin		Krogan		Wiphi	
	No. of nodes	No. of edges	No. of nodes	No. of edges	No. of nodes	No. of edges	No. of nodes	No. of edges
0	4928	17,201	1855	7669	2674	7075	5953	49,607
0.1	4928	17,185	1855	7669	2674	7075	5953	49,603
0.2	4927	17,133	1855	7669	2674	7067	5953	49,554
0.3	4923	17,026	1855	7665	2674	7042	5953	49,391
0.4	4892	16,793	1855	7654	2662	6960	5946	49,005
0.5	4771	16,356	1825	7587	2585	6782	5895	48,380
0.6	4673	15,984	1776	7501	2491	6608	5847	47,628
0.7	4661	15,719	1776	7472	2489	6538	5845	46,687
0.8	4661	15,480	1774	7424	2489	6479	5814	44,552
0.9	4462	14,711	1582	7015	2356	6275	4884	35,647

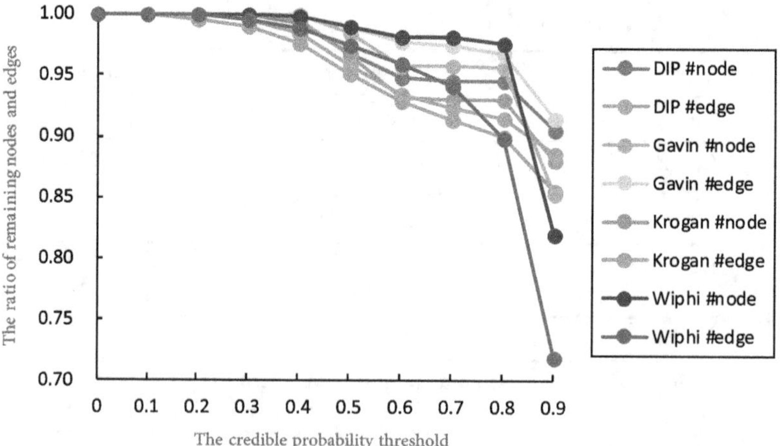

Fig. 4. The ratios of number of nodes/number of edges in the reconstructed PINs over those of the corresponding original PINs.

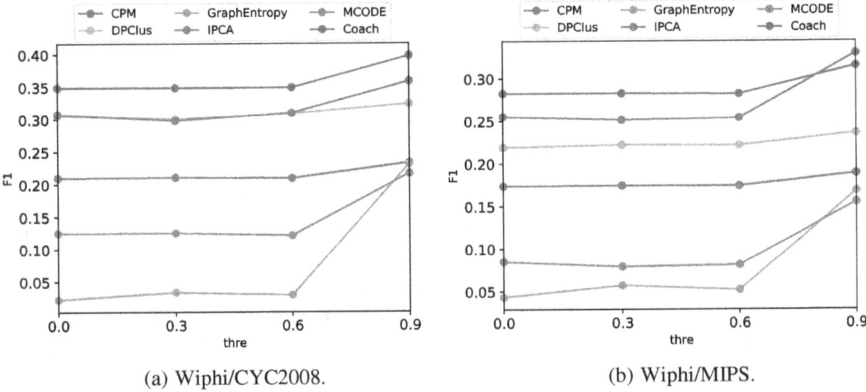

Fig. 5. Detection performances comparison for different threshold values. We change the threshold value while keeping the other parameters as in Table 4.

in [0,0.3), only a very small number of nodes/edges are removed, and when the threshold falls in [0.3,0.8), the filtering effect is still not very obvious. However, when the threshold reaches and surpasses 0.8, quite a lot of "noisy" interactions are removed. Thus, in the following experiments, we set the threshold to 0.9.

We also present the detection performances in Fig. 5 for different threshold values. As far as $F1$ is considered, we can also see that on the Wiphi dataset $F1$ reaches the maximum when the threshold is 0.9.

4.5. *Performance Comparison: The Reconstructed PINs Versus the Original PINs*

With the parameter settings in Table 4, we conduct complex detection on the original PINs and the reconstructed PINs, and the results are presented in Tables 6 and 7. Here, we use the *performance increase rate* (Δ) to measure the effect of our denoising method, which is defined as the ratio of the difference between the $F1$ of a reconstructed PIN and the F1 of the corresponding original PIN over the original $F1$.

From Table 6, we can see that in 39 out of 48 yeast cases, our method promotes detection performance from 0.1% to 656%. As for the remaining nine cases, seven happen on the dataset Krogan and two happen on the dataset Gavin (DPClus with MIPS as the benchmark) and dataset DIP (Coach with CYC2008 as the benchmark), respectively. In these nine cases, $F1$ decreases from 0.3% to 3%.

Table 6. Performance comparison between original PINs and the reconstructed PINs: Yeast datasets.

Methods	Datasets	Original PINs			Reconstructed PINs			
		Recall	Precision	$F1$	Recall	Precision	$F1$	Δ (%)
(a) Benchmark: MIPS								
CPM	DIP	0.6887	0.1243	0.2106	0.6745	0.1305	0.2187	3.83↑
	Gavin	0.5377	0.1696	0.2578	0.5094	0.1859	0.2724	5.66↑
	Krogan	0.4481	0.2096	0.2856	0.4528	0.2005	0.2779	−2.69↓
	Wiphi	0.8255	0.0972	0.1739	0.783	0.1073	0.1887	8.48↑
Coach	DIP	0.6462	0.2351	0.3447	0.6415	0.2438	0.3533	2.50↑
	Gavin	0.5047	0.2821	0.3619	0.4811	0.2925	0.3638	0.53↑
	Krogan	0.434	0.3496	0.3872	0.4292	0.3506	0.386	−0.32↓
	Wiphi	0.7358	0.1788	0.2877	0.6934	0.2038	0.315	9.50↑
DPClus	DIP	0.5425	0.2425	0.3352	0.5613	0.2415	0.3377	0.75↑
	Gavin	0.4953	0.2695	0.3491	0.467	0.2771	0.3478	−0.38↓
	Krogan	0.434	0.3081	0.3603	0.4292	0.2989	0.3524	−2.20↓
	Wiphi	0.6038	0.132	0.2167	0.6085	0.1463	0.2359	8.90↑
GraphEntropy	DIP	0.1274	0.1184	0.1227	0.1745	0.1333	0.1512	**23.21↑**
	Gavin	0.4292	0.266	0.3284	0.4198	0.3077	0.3551	8.13↑
	Krogan	0.3915	0.2141	0.2768	0.3962	0.2322	0.2928	5.79↑
	Wiphi	0.033	0.1279	0.0525	0.2594	0.1243	0.1681	**220.21↑**
IPCA	DIP	0.7453	0.1231	0.2113	0.75	0.1311	0.2232	5.61↑
	Gavin	0.533	0.2281	0.3195	0.5	0.2485	0.332	3.89↑
	Krogan	0.5802	0.2299	0.3293	0.5755	0.2469	0.3455	4.93↑
	Wiphi	0.6887	0.1569	0.2556	0.7123	0.2144	0.3296	**28.94↑**
MCODE	DIP	0.1887	0.4407	0.2642	0.1981	0.4516	0.2754	4.23↑
	Gavin	0.3679	0.3615	0.3647	0.3632	0.377	0.37	1.45↑
	Krogan	0.2689	0.3944	0.3197	0.2689	0.4286	0.3304	3.34↑
	Wiphi	0.0755	0.0694	0.0723	0.1792	0.1368	0.1552	**114.56↑**
(b) Benchmark: CYC2008								
CPM	DIP	0.715	0.1428	0.238	0.6891	0.1536	0.2512	5.53↑
	Gavin	0.5389	0.2232	0.3156	0.5337	0.239	0.3302	4.60↑
	Krogan	0.5233	0.3138	0.3924	0.5078	0.3047	0.3809	−2.93↓
	Wiphi	0.943	0.1161	0.2068	0.8653	0.1347	0.2331	12.72↑
Coach	DIP	0.6477	0.2813	0.3922	0.6166	0.2782	0.3834	−2.23↓
	Gavin	0.5078	0.3547	0.4177	0.4974	0.3761	0.4283	2.55↑
	Krogan	0.5181	0.4928	0.5052	0.4922	0.4848	0.4885	−3.31↓
	Wiphi	0.8187	0.2197	0.3464	0.7927	0.2654	0.3977	**14.79↑**
DPClus	DIP	0.5596	0.3144	0.4026	0.5389	0.322	0.4031	0.13↑
	Gavin	0.5078	0.3672	0.4262	0.5026	0.3939	0.4417	3.64↑
	Krogan	0.5026	0.5	0.5013	0.4819	0.5	0.4908	−2.10↓
	Wiphi	0.6788	0.1836	0.289	0.6788	0.2125	0.3237	12.00↑

Table 6. (*Continued*)

Methods	Datasets	Original PINs			Reconstructed PINs			
		Recall	Precision	$F1$	Recall	Precision	$F1$	Δ (%)
GraphEntropy	DIP	0.1451	0.1429	0.144	0.2073	0.1789	0.1921	**33.42**↑
	Gavin	0.4508	0.4043	0.4263	0.4404	0.4453	0.4429	3.90↑
	Krogan	0.4145	0.3364	0.3714	0.399	0.3591	0.378	1.78↑
	Wiphi	0.0207	0.0581	0.0306	0.3264	0.1788	0.231	**656.01**↑
IPCA	DIP	0.7772	0.161	0.2667	0.7824	0.1724	0.2826	5.94↑
	Gavin	0.544	0.2568	0.3489	0.5233	0.2776	0.3628	3.97↑
	Krogan	0.6114	0.2611	0.3659	0.601	0.2743	0.3767	2.96↑
	Wiphi	0.8031	0.1865	0.3027	0.7409	0.2366	0.3587	**18.50**↑
MCODE	DIP	0.1347	0.4407	0.2063	0.1503	0.4677	0.2275	**10.23**↑
	Gavin	0.3731	0.5	0.4273	0.3731	0.541	0.4416	3.35↑
	Krogan	0.2539	0.6338	0.3625	0.2435	0.6286	0.351	−3.17↓
	Wiphi	0.1192	0.1458	0.1312	0.2228	0.2105	0.2165	**650.60**↑

Table 7 exhibits the results of our denoising method on human PPI datasets. Among all the 18 cases, 12 of them gain better performance from 0.18% to 9.11%; in the remaining six cases, $F1$ decreases from 0.55% to 4.21%.

4.6. *Comparison with Existing Denoising Methods*

Here we compare our method with RWS and RedNemo, two typical existing PPI denoising methods based on PIN topology. Both RedNemo and RWS denoise PINs by removing spurious interactions and adding new interactions so that the total numbers of proteins and interactions remain invariant. Concretely, we first compare the $F1$ scores of complexes detection methods on reconstructed PINs by the above three methods, plus the original PINs. Next, we classify the reference protein complexes into separate categories based on predefined SIZE, EXT and DENS boundaries, separate recall scores between the classified ground truth and predictions are provided subsequently. Finally, we analyze the time complexity of the mentioned three denoising approaches.

4.6.1. *General F1 Performance*

As shown in Fig. 6, on the yeast dataset Wiphi, our method ranks first in all 12 cases; on the yeast dataset Krogan, our method outperforms the others in eight of the 12 cases; on the yeast datasets Gavin and DIP, our

Table 7. Performance comparison between original PINs and reconstructed PINs: Human datasets.

Methods	Datasets	Original PINs			Reconstructed PINs			Δ (%)
		Recall	Precision	$F1$	Recall	Precision	$F1$	
(a) Benchmark: REACT								
CPM	Reactome	0.6584	0.8356	0.7365	0.6512	0.8368	0.7324	−0.55↓
Coach	Reactome	0.4916	0.7771	0.6022	0.5249	0.8181	0.6395	**6.19↑**
DPClus	Reactome	0.4572	0.7607	0.5711	0.4498	0.7903	0.5733	0.39↑
GraphEntropy	Reactome	0.3007	0.6636	0.4139	0.3109	0.6847	0.4276	3.32↑
IPCA	Reactome	0.5773	0.6974	0.6317	0.5704	0.7217	0.6372	0.87↑
MCODE	Reactome	0.1632	0.8094	0.2717	0.1559	0.7867	0.2602	−4.21↓
(b) Benchmark: REACT_uniprotkb								
CPM	Reactomekb	0.6831	0.7747	0.7261	0.6700	0.7824	0.7218	−0.58↓
Coach	Reactomekb	0.5182	0.7535	0.6141	0.5573	0.7775	0.6492	**5.72↑**
DPClus	Reactomekb	0.4903	0.7635	0.5971	0.4842	0.7638	0.5927	−0.75↓
GraphEntropy	Reactomekb	0.3665	0.6859	0.4777	0.3492	0.6804	0.4615	−3.38↓
IPCA	Reactomekb	0.6272	0.7344	0.6766	0.6165	0.7537	0.6782	0.24↑
MCODE	Reactomekb	0.1946	0.7394	0.3081	0.2048	0.7601	0.3227	4.75↑
(c) Benchmark: CORE_COMPLEX_human								
CPM	Reactomekb	0.4656	0.3826	0.4200	0.4908	0.3865	0.4325	2.96↑
Coach	Reactomekb	0.2865	0.2978	0.2920	0.3012	0.3245	0.3124	6.98↑
DPClus	Reactomekb	0.2625	0.2825	0.2722	0.2583	0.2797	0.2686	−1.33↓
GraphEntropy	Reactomekb	0.1914	0.2118	0.2011	0.1785	0.2311	0.2014	0.18↑
IPCA	Reactomekb	0.3128	0.2500	0.2779	0.3006	0.2669	0.2828	1.75↑
MCODE	Reactomekb	0.1018	0.2736	0.1484	0.1123	0.2905	0.1620	**9.11↑**

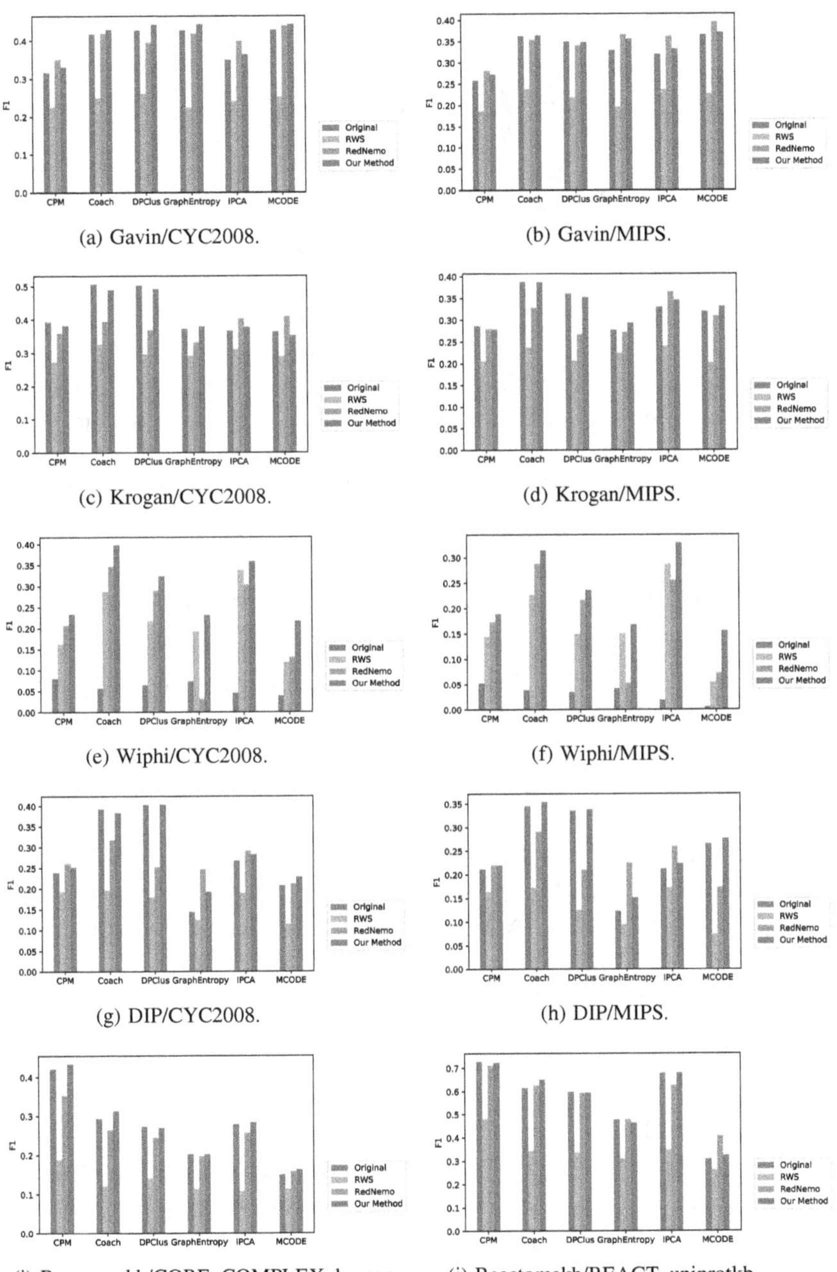

(a) Gavin/CYC2008.

(b) Gavin/MIPS.

(c) Krogan/CYC2008.

(d) Krogan/MIPS.

(e) Wiphi/CYC2008.

(f) Wiphi/MIPS.

(g) DIP/CYC2008.

(h) DIP/MIPS.

(i) Reactomekb/CORE_COMPLEX_human.

(j) Reactomekb/REACT_uniprotkb.

Fig. 6. Performance ($F1$) comparison with RWS and RedNemo.

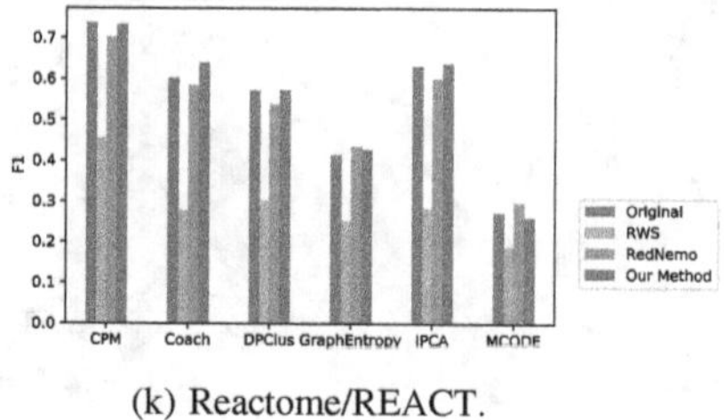

(k) Reactome/REACT.

Fig. 6. (*Continued*)

method achieves the best performance in six of the 12 cases. On the human dataset Reactome, our method ranks first in all six cases; on the human dataset Reactomekb, our method ranks first in seven of the 12 cases. In summary, in 45 out of 66 cases, our method outperforms the other two methods.

4.6.2. *Detailed Recall Performance on Varied Complexes Distributions*

To extensively and fairly show the potential of different denoising methods, i.e. on varied distributions with respect to SIZE, EXT and DENS of reference complexes, we select the relatively large yeast PPI dataset Wiphi and the human PPI dataset Reactome as our source PINs. The real reference complexes are from CYC2008 and REACT. All EXT and DENS values are computed based on the original PINs for a fair comparison.

Figures 7(a)–7(c) show the grouped *recall* results on CYC2008 with varied $\text{Rate}_{\text{match}}$ value of 0.2, 0.5 and 1.0. From them, we can see that (1) our proposed method achieves better performance than the other compared denoising methods, which can be easily seen when $\text{Rate}_{\text{match}}$ increases, especially on the row GraphEntropy and row MCODE. (2) The complexes that fall into the regions of high DENS and low EXT can be easily detected, as there is always a peak in almost every subfigure. (3) Detecting complexes with low DENS is a challenging task, because the low-density regions (denoted by the first two bars in each subfigure) are usually blank in a great quantity. (4) High EXT prohibits the predicting possibility from reaching a high rate, observing that real complexes in high EXT intervals have a lower recall, compared with those in the low EXT intervals.

On the human complex dataset REACT, as shown in Figs. 7(d)–7(f), there are more illustrations: (1) Our method still can follow up the characteristics of original PINs, instead, RedNemo performs worse when

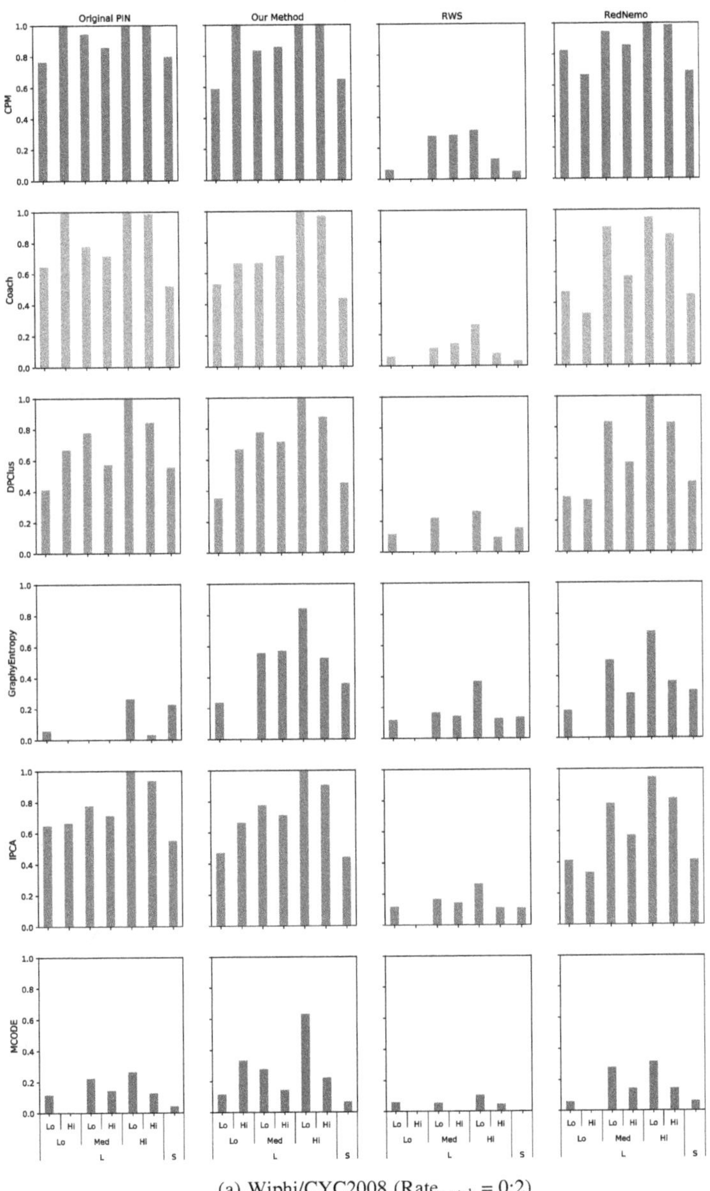

(a) Wiphi/CYC2008 (Rate$_{match}$ = 0:2)

Fig. 7. Performance (recall) comparison among original PINs and reconstructed PINs with different complex distributions. In Fig. 7, the labels of x-axis, from top, center to bottom, are EXT, DENS, and SIZE, separately.

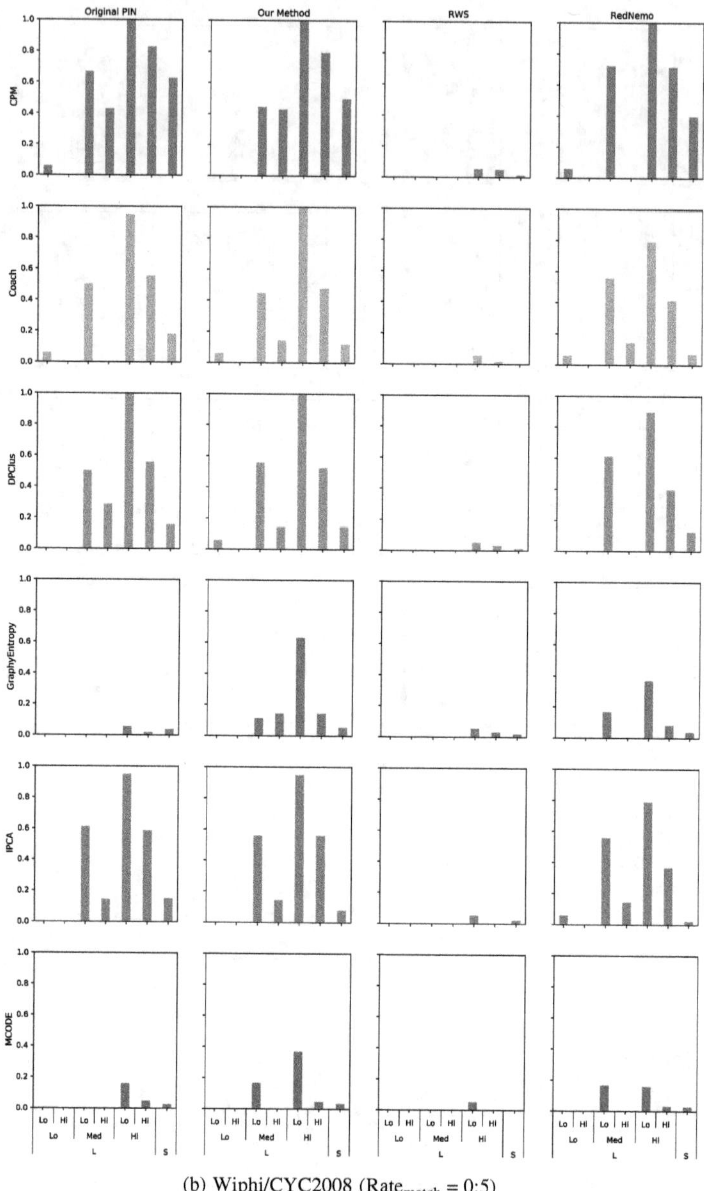

(b) Wiphi/CYC2008 $(\text{Rate}_{\text{match}} = 0:5)$

Fig. 7. (*Continued*)

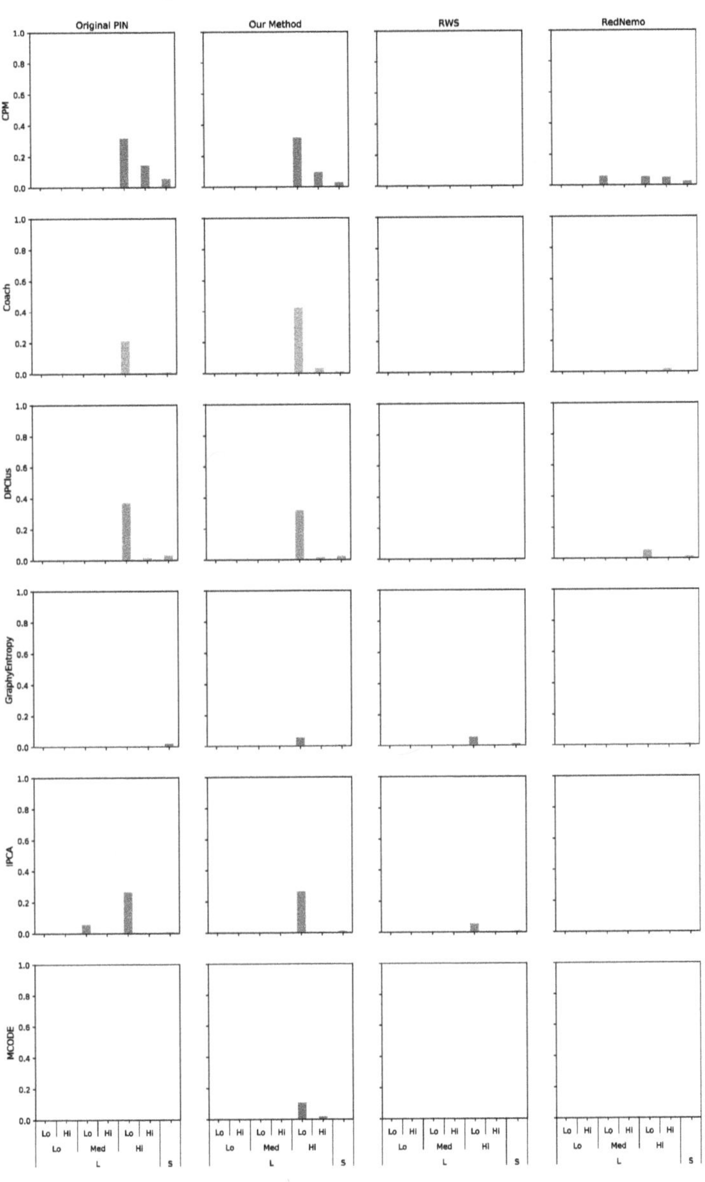

(c) Wiphi/CYC2008 ($\text{Rate}_{match} = 1:0$)

Fig. 7. (*Continued*)

Computational Approaches in Molecular Interaction Analysis

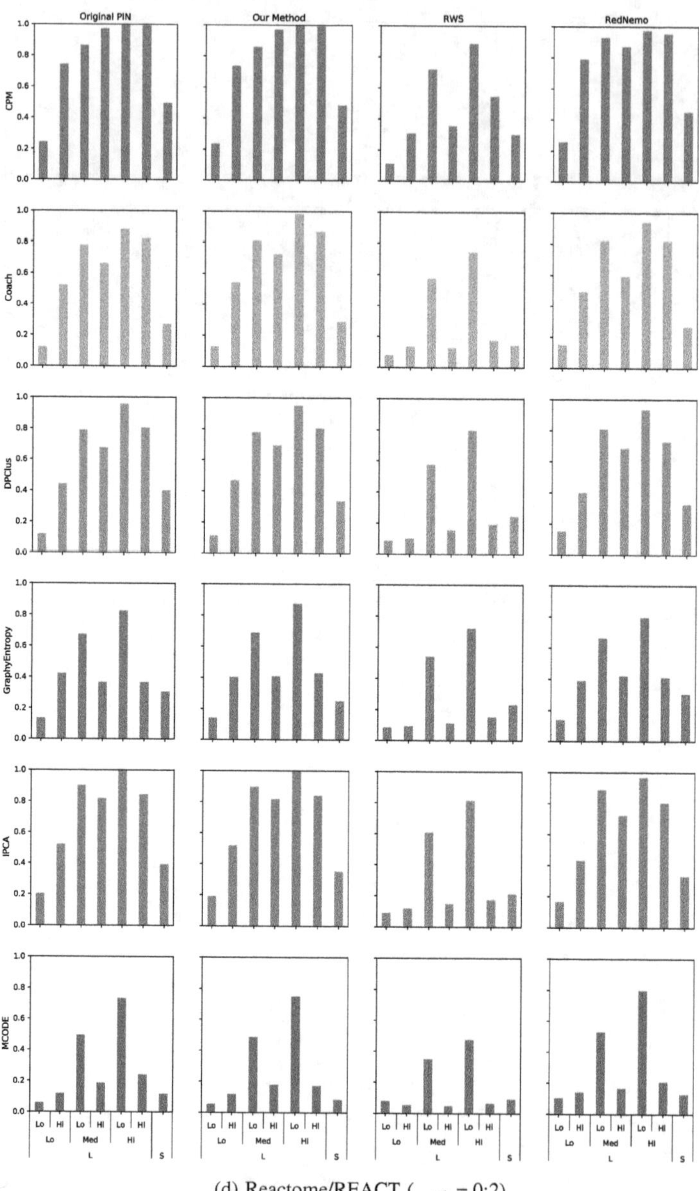

(d) Reactome/REACT ($_{match} = 0:2$)

Fig. 7. (*Continued*)

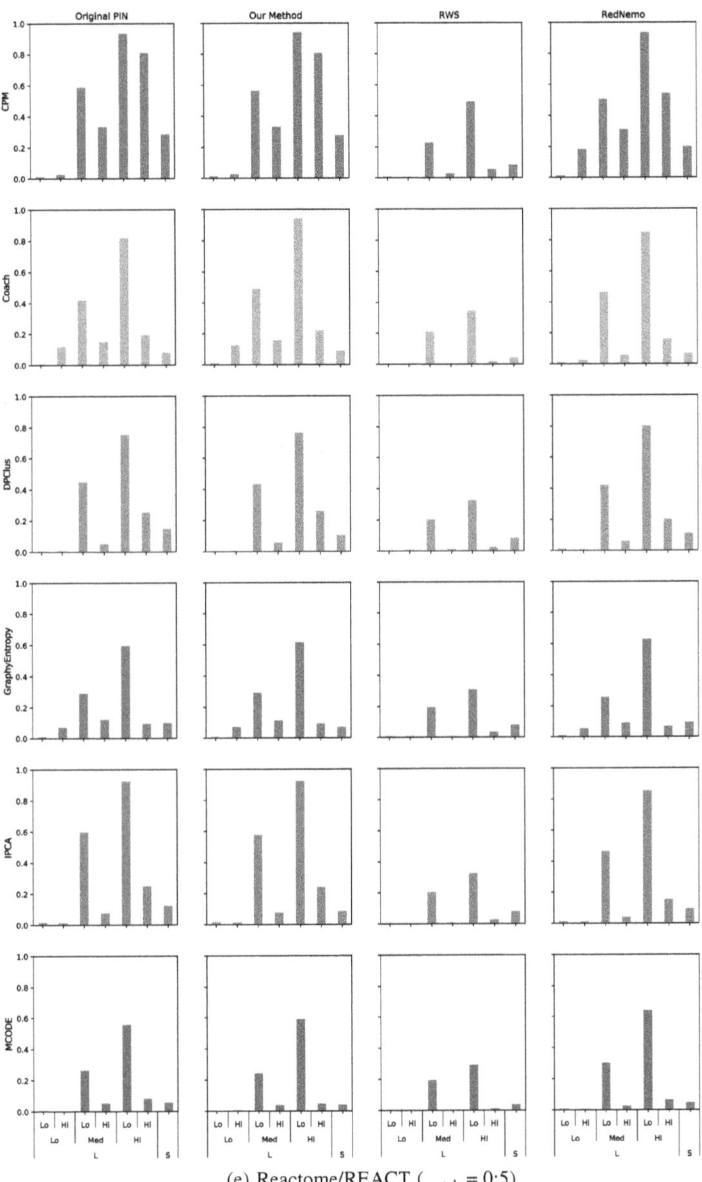

(e) Reactome/REACT $(_{match} = 0:5)$

Fig. 7. (*Continued*)

 Computational Approaches in Molecular Interaction Analysis

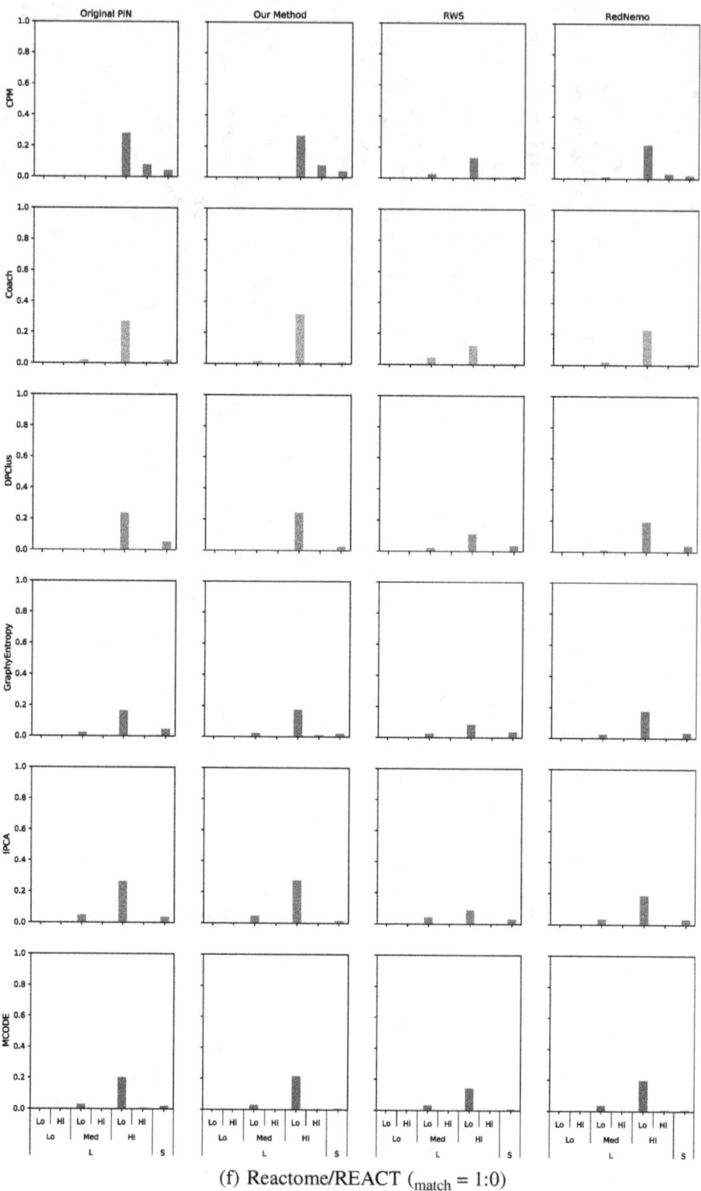

(f) Reactome/REACT ($_{match} = 1:0$)

Fig. 7. (*Continued*)

given a strict $\text{Rate}_{\text{match}}$ (see CPM row while $\text{Rate}_{\text{math}}$ is set to 1.0). (2) Besides high-density complexes, complexes with medium DENS and low EXT also can be predicted well (see the regions in medium DENS and low EXT).

4.6.3. *Time Complexity*

As for the time consumption, RWS and RedNemo have time complexity of $O(|V| \log |V|)$ ($|V|$ indicates the number of nodes in the PIN). Owing to the first-order approximation by Chebyshev polynomials, the time complexity of our method is $O(C|V|)$ where the constant C represents calculating the complexity of signal channels, feature maps and GCN neurons.

5. Conclusion

In this work, we propose a novel method to detect protein complexes by denoising false positive interactions in PINs based on the variational graph auto-encoder. Our experimental results show that with the reconstructed PINs, we can get better complex detection performance than with the original PINs. In most cases, performance improvement is more than 5%. Especially, while detecting complexes using GraphEntropy on a relatively large PPI dataset Wiphi, we obtain up to 220% (against the benchmark dataset MIPS) and 656% (against the benchmark dataset CYC2008) performance improvements, respectively. And compared with the two existing PPI denoising methods RedNemo and RWS, our method also achieves better performance.

In the future, to further boost complex detection performance, we plan to embed PINs with additional protein features, and design hybrid embedding models to evaluate the confidence of interactions. Furthermore, we will consider combining generative adversarial networks with graph auto-encoder to obtain better embeddings of PINs.

Acknowledgments

This work was supported by the National Natural Science Foundation of China (NSFC) (Grant No. 61772367) and Shanghai Municipal Commission of Economy and Informatization (Grant No. 18XI-05).

References

1. Enright AJ, Van Dongen S, Ouzounis CA, An efficient algorithm for large-scale detection of protein families, *Nucleic Acids Res* **30**(7):1575–1584, 2002.
2. King AD, Pržulj N, Jurisica I, Protein complex prediction via cost-based clustering, *Bioinformatics* **20**(17):3013–3020, 2004.
3. Adamcsek B, Palla G, Farkas IJ, Derényi I, Vicsek T, Cfinder: Locating cliques and overlapping modules in biological networks, *Bioinformatics* **22**(8):1021–1023, 2006.
4. Bader GD, Hogue CW, An automated method for finding molecular complexes in large protein interaction networks, *BMC Bioinformatics* **4**(1):2, 2003.
5. Spirin V, Mirny LA, Protein complexes and functional modules in molecular networks, *Proc Natl Acad Sci* **100**(21):12123–12128, 2003.
6. Leung HC, Xiang Q, Yiu SM, Chin FY, Predicting protein complexes from PPI data: A core-attachment approach, *J Comput Biol* **16**(2):133–144, 2009.
7. Wu M, Li X, Kwoh CK, Ng SK, A core-attachment based method to detect protein complexes in PPI networks, *BMC Bioinformatics* **10**:169, 2009.
8. Chua HN, Ning K, Sung WK, Leong HW, Wong L, Using indirect protein–protein interactions for protein complex prediction, *J Bioinform Comput Biol* **6**(3):435–466, 2008.
9. Ma CY, Chen YPP, Berger B, Liao CS, Identification of protein complexes by integrating multiple alignment of protein interaction networks, *Bioinformatics* **33**(11):1681–1688, 2017.
10. Ramadan E, Naef A, Ahmed M, Protein complexes predictions within protein interaction networks using genetic algorithms, *BMC Bioinformatics* **17**(7):269, 2016.
11. Theofilatos K, Pavlopoulou N, Papasavvas C, Likothanassis S, Dimitrakopoulos C, Georgopoulos E, Moschopoulos C, Mavroudi S, Predicting protein complexes from weighted protein–protein interaction graphs with a novel unsupervised methodology: Evolutionary enhanced Markov clustering, *Artif Intell Med* **63**(3):181–9, 2015.
12. Maraziotis IA, Dimitrakopoulou K, Bezerianos A, Growing functional modules from a seed protein via integration of protein interaction and gene expression data, *BMC Bioinformatics* **8**:408, 2007.
13. Ou-Yang L, Yan H, Zhang XF, A multi-network clustering method for detecting protein complexes from multiple heterogeneous networks, *BMC Bioinformatics* **18**(13):463, 2017.
14. Wang J, Peng X, Li M, Pan Y, Construction and application of dynamic protein interaction network based on time course gene expression data, *Proteomics* **13**(2):301–312, 2013.

15. Shi Y, Yao H, Guan J, Zhou S, CPredictor 4.0: Effectively detecting protein complexes in weighted dynamic PPI networks, *Int J Data Min Bioinform* **20**(4):303–319, 2018.

16. Xu B, Guan J, From function to interaction: A new paradigm for accurately predicting protein complexes based on protein-to-protein interaction networks, *IEEE/ACM Trans Comput Biol Bioinform* **11**(4):616–627, 2014.

17. Xu B, Wang Y, Wang Z, Zhou J, Zhou S, Guan J, An effective approach to detecting both small and large complexes from protein-protein interaction networks, *BMC Bioinformatics* **18**(12):419, 2017.

18. Xu Y, Zhou J, Zhou S, Guan J, CPredictor3.0: Detecting protein complexes from ppi networks with expression data and functional annotations, *BMC Syst Biol* **11**(7):135, 2017.

19. Yao H, Shi Y, Guan J, Zhou S, Accurately detecting protein complexes by graph embedding and combining functions with interactions, *IEEE/ACM Trans Comput Biol Bioinform*, doi:10.1109/TCBB.2019.2897769, 2019.

20. Yu FY, Yang ZH, Hu XH, Sun YY, Lin HF, Wang J, Protein complex detection in PPI networks based on data integration and supervised learning method, *BMC Bioinformatics* **16**(Suppl 12):S3, 2015.

21. Yu FY, Yang ZH, Tang N, Lin HF, Wang J, Yang ZW, Predicting protein complex in protein interaction network: A supervised learning based method, *BMC Syst Biol* **8**(Suppl 3):S4, 2014.

22. Tian F, Gao B, Cui Q, Chen E, Liu TY, Learning deep representations for graph clustering, *Proc Twenty-Eight AAAI Conf Artificial Intelligence*, pp. 1293–1299, 2014.

23. Perozzi B, Al-Rfou R, Skiena S, DeepWalk: Online learning of social representations, *Proc 20th ACM SIGKDD Int Conf Knowledge Discovery and Data Mining*, pp. 701–710, 2014.

24. Grover A, Leskovec J, node2vec: Scalable feature learning for networks, *Proc 22nd ACM SIGKDD Int Conf Knowledge Discovery and Data Mining*, pp. 855–864, 2016.

25. Tang J, Qu M, Wang M, Zhang M, Yan J, Mei Q, LINE: Large-scale information network embedding, *Proc 24th Int Conf World Wide Web*, pp. 1067–1077, 2015.

26. Wang D, Cui P, Zhu W, Structural deep network embedding, *Proc 22nd ACM SIGKDD Int Conf Knowledge Discovery and Data Mining*, pp. 1225–1234, 2016.

27. Gavin AC *et al.*, Proteome survey reveals modularity of the yeast cell machinery, *Nature* **440**(7084):631–636, 2006.

28. Xenarios I, Salwinski L, Duan XJ, Higney P, Kim SM, Eisenberg D, DIP, the database of interacting proteins: A research tool for studying cellular networks of protein interactions, *Nucleic Acids Res* **30**(1):303–305, 2002.

29. Krogan NJ *et al.*, Global landscape of protein complexes in the yeast *Saccharomyces cerevisiae*, *Nature* **440**(7084):637–643, 2006.

30. Kiemer L, Costa S, Ueffing M, Cesareni G, WI-PHI: A weighted yeast interactome enriched for direct physical interactions, *Proteomics* **7**(6):932–943, 2007.

31. Pu S, Wong J, Turner B, Cho E, Wodak SJ, Up-to-date catalogues of yeast protein complexes, *Nucleic Acids Res* **37**(3):825–831, 2008.

32. Mewes HW *et al.*, MIPS: Analysis and annotation of proteins from whole genomes, *Nucleic Acids Res* **32**(suppl_1):D41–D44, 2004.

33. Croft D *et al.*, Reactome: A database of reactions, pathways and biological processes, *Nucleic Acids Res* **39**(suppl_1):D691–D697, 2010.

34. Giurgiu M, Reinhard J, Brauner B, Dunger-Kaltenbach I, Fobo G, Frishman G, Montrone C, Ruepp A, CORUM: The comprehensive resource of mammalian protein complexes — 2019, *Nucleic Acids Res* **47**(D1):D559–D563, 2019.

35. Yong CH, Wong L, From the static interactome to dynamic protein complexes: Three challenges, *J Bioinform Comput Biol* **13**(2):1571001, 2015.

36. Kuchaiev O, Rašajski M, Higham DJ, Pržulj N, Geometric de-noising of protein-protein interaction networks, *PLoS Comput Biol* **5**(8):e1000454, 2009.

37. Lei C, Ruan J, A novel link prediction algorithm for reconstructing protein–protein interaction networks by topological similarity, *Bioinformatics* **29**(3):355–364, 2012.

38. Alkan F, Erten C, RedNemo: Topology-based PPI network reconstruction via repeated diffusion with neighborhood modifications, *Bioinformatics* **33**(4):537–544, 2016.

39. Cui P, Wang X, Pei J, Zhu W, A survey on network embedding, *IEEE Trans Know Data Eng* **31**(5):833–852, 2019.

40. Hamilton WL, Ying R, Leskovec J, Representation learning on graphs: Methods and applications, preprint, arXiv:1709.05584 [cs.SI], 2017.

41. Kipf TN, Welling M, Variational graph auto-encoders, *Proc NIPS Workshop on Bayesian Deep Learning*, arXiv:1611.07308 [stat.ML], 2016.

42. Hammond DK, Vandergheynst P, Gribonval R, Wavelets on graphs via spectral graph theory, preprint, arXiv:0912.3848 [math.FA], 2009.

43. Wang C, Pan S, Long G, Zhu X, Jiang J, MGAE: Marginalized graph autoencoder for graph clustering, *Proc 2017 ACM Conf Information and Knowledge Management*, pp. 889–898, 2017.

44. Pan S, Hu R, Long G, Jiang J, Yao L, Zhang C, Adversarially regularized graph autoencoder for graph embedding, *Proc Twenty Seventh Int Joint Conf Artificial Intelligence*, pp. 2609–2615, 2018.

45. Defferrard M, Bresson X, Vandergheynst P, Convolutional neural networks on graphs with fast localized spectral filtering, in Lee DD, Sugiyama M,

Luxburg UV, Guyon I, Garnett R (eds.), *Advances in Neural Information Processing Systems*, *29*, Curran Associates Inc., Red Hook, pp. 3844–3852, 2016.

46. Kipf TN, Welling M, Semi-supervised classification with graph convolutional networks, *Proc Int Conf Learning Representations (ICLR)*, arXiv:1609.02907 [cs.LG], 2017.

47. Vincent P, Larochelle H, Bengio Y, Manzagol PA, Extracting and composing robust features with denoising autoencoders, *Proc 25th Int Conf Machine Learning*, pp. 1096–1103, 2008.

48. Altaf-Ul-Amin M, Shinbo Y, Mihara K, Kurokawa K, Kanaya S, Development and implementation of an algorithm for detection of protein complexes in large interaction networks, *BMC Bioinformatics* **7**(1):207, 2006.

49. Price T, PeñaFI 3rd, Cho YR, Survey: Enhancing protein complex prediction in PPI networks with GO similarity weighting, *Interdiscip Sci, Comput Life Sci* **5**(3):196–210, 2013.

50. Li M, Chen J, Wang J, Hu B, Chen G, Modifying the DPClus algorithm for identifying protein complexes based on new topological structures, *BMC Bioinformatics* **9**(1):398, 2008.

51. Kenley EC, Cho YR, Detecting protein complexes and functional modules from protein interaction networks: A graph entropy approach, *Proteomics* **11**(19):3835–3844, 2011.

Heng Yao received his Bachelor's degree from Tongji University, Shanghai, China, in 2016, and is currently working toward the Ph.D. degree at the Department of Computer Science and Technology, Tongji University. His research interests include network representation learning and bioinformatics.

Jihong Guan received the Bachelor's degree from Huazhong Normal University in 1991, the Master's degree from the Wuhan Technical University of Surveying and Mapping (merged into Wuhan University in 2000) in 1998 and the Ph.D. degree from Wuhan University in 2002. She is currently a Professor at the Department of Computer Science and Technology, Tongji University, Shanghai, China. Her research interests include databases, data mining, distributed computing, bioinformatics and geographic information systems. She has extensively published more than 300 papers in domestic and international journals (including Nature Communications, IEEE TKDE/TITS/TSC/TGRS/TCBB, NAR and Bioinformatics) and conferences (including AAAI, ICDE, VLDB, SIGIR, RECOMB and DASFAA).

Tianying Liu received the Bachelor's degree from Tongji University, Shanghai, China, in 2018. He is currently working toward the Ph.D. degree at the Department of Computer Science and Technology, Tongji University. His research interests include computer vision and deep learning.

CHAPTER 3

Identifying Protein Complexes from Protein–Protein Interaction Networks Based on the Gene Expression Profile and Core-Attachment Approach[a]

Soheir Noori[*,†,§], Nabeel Al-A'Araji[‡,¶] and Eman Al-Shamery[*,∥]

*Software Department, University of Babylon
Babylon, Hillah, Iraq*

*†Computer Science Department
University of Kerbala, Babylon, Hillah, Iraq*

‡Ministry of Higher Education, Baghdad, Iraq
§soheirn.sw.hdr@student.uobabylon.edu.iq;
Soheir.noori@uokerbala.edu.iq
¶nhkaghed@itnet.uobabylon.edu.iq
∥emanalshamery@itnet.uobabylon.edu.iq

Defining protein complexes in the cell is important for learning about cellular processes mechanisms as they perform many of the molecular functions in these processes. Most of the proposed algorithms predict a complex as a dense area in a Protein–Protein Interaction (PPI) network. Others, on the other hand, weight the network using gene expression or geneontology (GO). These approaches, however, eliminate the proteins and their edges that offer no gene expression data. This can lead to the loss of important topological relations. Therefore, in this study, a method based on the Gene Expression and Core-Attachment (GECA) approach was proposed for addressing these limitations. GECA is a new technique to identify core proteins using common neighbor techniques and biological information. Moreover, GECA improves the attachment

---

[a]This article was previously published in *Journal of Bioinformatics and Computational Biology*. Vol: 19, No. 3 (2021), 2150009 (17 pages).

technique by adding the proteins that have low closeness but high similarity to the gene expression of the core proteins. GECA has been compared with several existing methods and proved in most datasets to be able to achieve the highest F-measure. The evaluation of complexes predicted by GECA shows high biological significance.

Keywords: Protein complex; protein–protein interaction; gene expression profile; core-attachment approach.

1. Background

Protein complexes are the key to understanding the mechanism and organization of cell processes since they take part in essential cellular functions.[1] These protein complexes are combinations of two or more proteins assembled from the proteins in the Protein–Protein Interaction Network (PPIN). During the past decade, scientific research has directed more attention to detecting protein complexes and developing numerous algorithms for dealing with them. Most of these algorithms predict a protein complex as a dense area in a PPIN. They do this either by using a density threshold or an objective density function such as MCODE[2] which uses graph theory to determine dense sub-graphs from a weighted PPIN in a greedy manner. DPClus[3] and IPCA[4] search for dense sub-graphs that have a small diameter, while other algorithms search for cliques. Researchers believe that a complete connected subgraph, which is not too small, corresponds to the core of a protein complex such as CFinder,[5] Maximal Cliques[6] or NCMine.[7] All these algorithms find cliques and then merge them depending on various criteria for identifying the dense sub-graphs that represent the protein complexes. However, these methods ignore many low-density complexes, even though 40% of the complexes in CYC2008,[8] MIPS,[9] and Aloy[10] have a density of less than 0.5.[11]

Gavin *et al.*[12] reported that protein complexes were organized as core proteins and attachment proteins. The core represents the proteins that strongly interact with each other, whereas the attachments mark the boundary of the core. Several methods based on this technique have also been proposed, such as CORE,[13] COACH,[14] and WCOACH,[15] which improved COACH by using Gene Ontology (GO) to weight the PPIN. These algorithms contain two phases, namely, identifying the complex core as a dense sub-graph, and then identifying the attachment proteins as those that interact with more than half the proteins in the core.

Nevertheless, most protein complex detection algorithms employ the topological properties of a graph or mix PPIN with other information, such as the gene expression that reveals the dynamic properties of the PPIN. Hunter[16] and WEC[17] demonstrated that a weighted PPIN by its gene expression profile (GEP) could improve the quality of protein complexes detection. Here WEC weighted the edge between two proteins by the Edge Clustering Coefficient (ECC) and the correlation between the proteins' gene expression, which made it possible to predict the protein complex accurately. CAG[18] used the core-attachment technique and identified the core as a functional unit by finding only the positive correlation between proteins in the neighborhood and iteratively remove nodes with the minimum degree to maintain the density of the core. They then attached the proteins to the core if they were connected with half of the core proteins. Most algorithms that used the gene expression and GO to weight PPIN, such as WCOACH, WEC, and CAG eliminate edges that cannot locate a biological information for any of their proteins, thereby ignoring important topological relations between proteins.

The paragraphs above have raised several issues. First, while algorithms based on density are effective in predicting complexes,[19] some complexes may have a density of less than 0.5.[11] Second, the elimination of edges from the PPIN when biological information about their proteins is absent may lead to bias in the results. Third, most core-attachment-based methods employ the same technique in adding attachment proteins, ignoring the dynamic properties of the PPIN. Therefore, in order to address these limitations, this paper proposes a new method of identifying protein complexes based on the Gene Expression profile (GEP) and Core-Attachment approach (GECA). The GECA proposes a new technique that both identifies core proteins using common neighbor techniques and GEP, and improves the attachment technique by adding proteins that have fewer connections but are more similar in gene expression with core proteins. GECA does not remove any edges from the PPIN as other algorithms do.

1.1. *Preliminaries*

Generally, the PPIN is represented as an undirected and unweighted graph $G(V, E)$ where the nodes (V) are proteins, and the edges (E) are the set of protein interactions. First, some terms for describing the method should be defined.

Definition 1. Protein neighbors: the neighbors of protein v are the set of proteins that interact directly with protein v and are denoted as $|Nv|$.

Definition 2. GEP is a $t \times n$ matrix of real values to t genes or proteins in n conditions or at n time points. The expression value of protein v can be represented as $Ge(v) = [g(v,1), g(v,2), \ldots, g(v,n)]$.

The expression values of two genes or proteins X and Y are represented as $X = [x_1, x_2, x_3, \ldots, x_n]$ and $Y = [y_1, y_2, y_3, \ldots, y_n]$. Pearson's Correlation Coefficient (PCC) as defined in Eq. (1) is used to measure the similarity between these two GEP patterns. PCC is the most successful way to measure the similarity of the co-expression between genes.[20]

$$\text{PCC}(X, Y) = \frac{\sum_{i=1}^{n}(x_i - x')(y_i - y')}{\sqrt{\sum_{i=1}^{n}(x_i - x')^2 \sum_{j=1}^{n}(y_i - y')^2}}. \tag{1}$$

PCC determines the similarity between two genes X and Y. x_i to x_n and y_i to y_n represented in the expression values of gene X and Y at n time points, x' and y' representing the mean of the expression value of X and Y genes. The value of PCC is between -1 and 1. 1 represents the perfect correlation between two genes in a positive direction, whereas -1 represents the perfect correlation but in a negative direction, and 0 represents no correlation between two genes.

In most clustering methods, ECC is used to determine the closeness between two nodes in a network. ECC is defined in Eq. (2) where Triangle (V_1, V_2) signifies the number of common neighbors between node V_1 and node V_2 divided by the minimum degree of the two nodes. In this paper, we employ a common neighbor (CN) technique to determine the closeness between the node and its direct neighbors (as defined in Eq. (3) below). CN, between node V_1 and its neighbor V_2, is determined by the number of nodes that directly interact with both of them divided by the square root of the product of the degrees of the nodes. Radicchi *et al.*[21] noted that ECC may not be suitable for PPIN because it is disassortative. CN is used in this paper to overcome this ECC limitation.

$$\text{ECC} = \frac{\text{Triangle}(V_1, V_2)}{\min\{\deg(V_1) - 1, \deg(V_2) - 1\}}, \tag{2}$$

$$\text{CN} = \frac{N_{V_1} \cap N_{V_2}}{\sqrt{d(V_1) * d(V_2)}}. \tag{3}$$

Furthermore, this paper integrates topological and dynamic properties of PPIN by using CN and PCC. Here CN captures the topological relation between any node and its neighbors, while PCC captures the dynamic properties of the proteins by using the gene expression over a time series. In addition, the edges of the network are weighted by the value of PCC that represents the degree of similarity between the gene expression patterns of the two proteins if the gene expression data is available for both proteins. Otherwise the weight is set to zero.

2. Method

In this study, protein complexes are predicted in three steps, namely, core construction, merging cores, and adding an attachment, described as follows:

Core construction: The proteins in the core of a protein complex constantly interact and the gene expression between them is highly correlated.[22] The procedure begins with a seed v as a preliminary core, where neighbors of the seed are added either if the CN $>= T_{core}$ (T_{core} is the threshold that determines the closeness between the seed and its neighbors), or if the PCC is greater than T_{ppc}, according to the correlation between the two proteins. The core contained the proteins that were either closely connected or correlated in gene expression. The core is accepted if its density is greater than the density threshold T_{dense}, which is set to 0.7 as in previous studies.[3,23] The use of PCC in the core allows the addition of peripheral protein that has an expressed relationship with the seed. The process of core identification is done by applying Algorithm 1.

Algorithm 1. Core_construction

Input: weighted PPIN by PCC (WG); T_{core}; T_{pcc}
Output: the set of Cores (CORES)
1: **For** v in WG **do** /* v is a vertex in WG */
2:     Temp_Core = Temp_Core union v /* add v to Temporary_Core */
3:     **For** u in Nv do /* u is neighbor of vertex v */
4:        **If** CN $(v, u) >= T_{core}$ or PCC $(v, u) >= T_{pcc}$ **then**
5:           Temp_Core = Temp_Core union u /* add u to Temp_Core */
6:     **end for**
7:     If Temp_Core not in CORES and len(Temp_Core) > 2 and density(Temp_Core) $>= T_{dense}$ then
8:        CORES = CORES union Temp_Core /* accept Temp_Core as a final Core */
9: **end for**

Merged cores: Algorithm 1 identified the cores which had many overlapping proteins. It was sufficient to merge highly overlapping cores since the density of the cores was equal to or greater than 0.7. When the overlapping scores (OS) between two cores, C_1 and C_2, as in Eq. (4), are greater than the $T_{\text{filtering}}$, which was set to 0.9, the two cores merge. When the core does not overlap with other cores, it is kept as a core in the list of cores.

$$OS(C_1, C_2) = \frac{|V_{C_1} \cap V_{C_2}|^2}{|V_{C_1}| \times |V_{C_2}|}. \tag{4}$$

Adding the attachment: Most of the core-attachment-based methods followed the CoAch[14] procedure mentioned above, in which a protein is added to the core if it has a direct connection with half of the core proteins. However, in actual complexes, the proteins in the attachment interact with fewer than half of the core proteins.[12] The process of adding proteins to the core is ineffective and requires further information that can be obtained from the gene expression, which provides a more meaningful biological result. In this study, however, the protein is added to the core so long as it meets one of the following two conditions:

(1) The closeness score[14] is greater than $T_{\text{closeness}}$ as given in Eq. (5), where N_u represents the direct neighborhoods of u and V_{core} represents the vertices in Core.

$$\text{Closeness}(\text{Core}, u) = \frac{|N_u| \cap |V_{\text{core}}|}{|V_{\text{core}}|} \geq T_{\text{closeness}}. \tag{5}$$

OR

(2) The interWeight Eq. (6) is equal to or greater than coreWeight Eq. (7) and node u has at least n connections with the core.

$$\text{interWeight}(u, \text{core}) = \frac{\sum_e w(e)}{|N_u| \cap |V_{\text{core}}|}, \tag{6}$$

$$\text{coreWeight}(\text{core}) = \frac{\sum_e w(e)}{|V_{\text{core}}|}, \tag{7}$$

where $\sum_e w(e)$ in Eq. (6) is the sum of weights of all interactions between protein u and the core proteins normalized by the number of proteins in the core that protein u interact with. While $\sum_e w(e)$ in Eq. (7) is the sum of all weights in cores normalized by the number of the core proteins.

Algorithm 2. Adding an attachment

Input: set of CORES
Output: set of PCs
1: **For** core in CORES **do**
2:     proteins_not_in_core = ∅ /* set of proteins that are not in Core */
3:     **For** protein in core **do**
4:         difference = $\{u | u \in N_{protein}, u$ not in core$\}$ /* the neighbors of protein that are not in Core*/
5:     **end for**
6:     **For** protein in difference **do**
7:         **If** (Closeness (core, protein) $>= T_{closeness}$) or (interWeight (protein, core) $>=$ coreWeight (core) and intersection ($N_{protein}$, core) $> N_{connection}$) **then**
8:             protein_not_in_core = protein_not_in_core union protein
9:     **end for**
10:     **If** len(protein_not_in_core) > 0 **then**
11:         **For** protein in protein_not_in_core **do**
12:             core = core union protein /* add attachment proteins to the Core */
13:         **end for**
14: **end for**

A protein is added if it highly interacts with the proteins in the core, or has a good similarity with them in the gene expression pattern. The aim here is to improve the attachment method in order to add proteins that interact with less than half of the core proteins but have similar gene expression patterns. Finally, the set of predicted complexes (PCs) is defined, as expressed in Algorithm 2.

2.1. *Evaluation Metrics*

2.1.1. *Recall, Precision and F-Measure*

GECA was compared with the other methods, using precision Eq. (8), recall Eq. (9) and F-measure Eq. (10) as evaluation criteria. Let $P = \{p_1, p_2, p_3, \ldots, p_n\}$ a set of PCs and $B = \{b_1, b_2, b_3, \ldots, b_n\}$ a set of actual complexes. The recall, precision and F-measure are defined as follows:

$$\text{Precision} = \frac{|\{p | p \in P, \exists b \in B, \alpha(b,p) \geq \theta\}|}{|P|}, \tag{8}$$

$$\text{Recall} = \frac{|\{b | b \in B, \exists p \in P, \alpha(b,p) \geq \theta\}|}{|B|}, \tag{9}$$

$$\text{F-measure} = \frac{2 \times \text{Precision} \times \text{Recall}}{\text{Precision} + \text{Recall}}. \tag{10}$$

In this paper, $\alpha(b,p)$ is either the OS between the actual and the PC, as used in all compared algorithms, or the Jaccard similarity (Jac-Sim) Eq. (11) between the actual and PC is based on that of Yong and Wong.[24] θ for OS was determined as 0.255 as in WEC and CAG, θ for Jac-Sim was 0.75 for yeast and 0.50 for human or other large complexes greater than 3 in size, while θ was equal to 1 for small complexes.

$$\text{Jac}-\text{Sim}(P,B) = \frac{|V_P \cap V_B|}{|V_P \cup V_B|}. \tag{11}$$

2.1.2. *Co-Localization Score and Gene Ontology (GO) Semantic Similarity Score*

The predicted complex might be valid, but might not match any actual complex since the standard set of actual complexes is not complete.[25] The co-localization score, Eq. (12), is allowed to modify the quality of PCs that do not match the actual complexes. The basic idea of the co-localization score is that the proteins in the complex should be found in the same cellular compartment[26] and more than likely to be involved in the same function.

$$\text{CO}-\text{localization score} = \frac{\sum_i \max_j N_{ij}}{\sum_i |C_I|}, \tag{12}$$

where N_{ij} is the number of proteins in the C_i complex that is assigned to j, the localization group, and $|C_i|$ is the number of proteins in C_i complex assigned to localization.

The GO annotation is used to evaluate the functional similarity between the proteins of the complex. The functional similarity between proteins is based on measuring the similarity in GO terms that annotate these proteins.[27] Accordingly, the greater the score of GO semantic similarity, the better the quality of the predicted complex is.

The Procope software tool is used for co-localization and GO semantic similarity scores[28] and the data used in the evaluation process is set to "default".

2.2. *Data Sources*

The datasets used in this paper comprise the same data used in the WEC and CAG algorithms. They are two datasets from yeast saccharomyces cerevisiae (Collins *et al.*[29] and Gavin *et al.*[12] with GEP from Tu *et al.*[30])

Table 1. Datasets details.

Dataset	
Collins	9074 interactions
Gavin	7669 interactions
Yeast GEP	9335 gene and 36-time points
Homo sapiens PPI network	37437 interactions
Homo sapiens GEP	54675 genes and 27-time points
MIPS	203 complexes
SGD	323 complexes
CORUM	1521 complexes

and the benchmark datasets taken from MIPS[31] and SGD.[32] Only high confidence interactions greater than five were considered. However, the GEP data did not cover 792 interactions in Collin and 152 interactions in Gavin. The PPIN of human is integrated from HPRD (the Human Protein Reference Database Release 9)[33] and HSN (Human Signalling Network)[34] The GEP of Homo sapiens[35] is downloaded from http://www.ncbi.nlm. nih.gov/sites/GDSbrowser?acc=GDS2604. However, the GEP did not cover 654 of PPIN interactions. The benchmark complexes of Homo sapiens have been obtained from CORUM.[36] Details of the datasets are presented in Table 1. The connections that are not covered by GEP are not removed.

2.3. *Analysis of Actual Complexes*

GEP has been applied on the actual complexes SGD, MIPS, and CORUM to analyze the correlation between the proteins present in the same complex. Thereafter, the complex has been stratified to the core and attachment. Most correlations between the complex proteins are positive, and the incidence of a negative correlation between them is low. The results are the same between the core proteins, the attachment proteins and, further, between the core and attachment proteins. File S1 describes the details concerning this analysis. Moreover, the data analysis made in this study leads to the conclusion that GECA depends on obtaining all the positive correlations as well as having a low negative correlation to add proteins to the core in the yeast data. In the human data, GECA uses only the low positive and low negative correlations to add proteins to the core.

2.4. *Threshold Values*

The experiment was conducted with different parameter values. As shown in Algorithm 1, there were two thresholds where the T_{core} determined the number of common neighbors between the seed and its neighbors, which used 0.2 in all data. T_{pcc} was kept at 0.1 for a positive correlation and equal to or greater than -0.5 for a negative correlation in the yeast data. Meanwhile, in the human data, the T_{pcc} was greater than 0.5 for a positive correlation and equal to or greater than -0.5 for a negative correlation. Whereas Algorithm 2 had two thresholds, $T_{\text{closeness}}$ was kept at 0.7 in the Gavin and Collins data and 0.9 in the human data in order to add only the protein that is closely connected with the core proteins. $N_{\text{connection}}$ was kept at 2 in the Collins and Gavin data and 3 in the human data to allow the addition of the protein with low connections to the core but high correlation to the gene expression pattern.

2.5. *Comparison with Other Methods*

The GECA algorithm is next compared to four methods. ClusterONE[37] starts from seed protein in the highest degree, then gradually adding and removing proteins to find a cohesive group of proteins which could be overlapped. WEC[17] and CAG,[18] previously mentioned in the introduction section, used the same datasets, but removed unweighted interactions between proteins. Finally, WCOACH[15] used GO to weight the PPIN and also remove the unweighted edges. Every parameter in all the algorithms was set to default or as the papers on them recommended. In addition, complexes with less than three proteins were ignored. Section 3 describes the effects of removing unweighted edges on the result, especially in the Collins data from which the highest ratio of edges were removed.

3. Results and Discussion

3.1. *F-Measure Results of Yeast and Human Data*

Gavin data: GECA predicted 538 complexes, 238 of which matched with actual MIPS complexes and 271 matched with SGD complexes.

The maximum predicted complex contained 43 proteins. When using MIPS as benchmark data, GECA achieved recall, precision, and F-measure values of 0.453, 0.442, and 0.448, respectively. GECA achieved the highest F-measure. The other methods (CAG, WEC, WCOACH, and ClusterONE) achieved the F-measure values of 0.442, 0.429, 0.408, and 0.314, respectively. When SGD was used as benchmark data, the F-measure value of GECA was 0.438, followed by CAG with 0.41, WEC with 0.409, WCOACH with 0.38 and ClusterONE having 0.35. GECA achieved the highest F-measure with respect to the Gavin data when using SGD and MIPS as benchmark complexes.

Collins data: GECA predicted 574 complexes, 302 of which matched with MIPS actual complexes and 292 matched with SGD actual complexes. The maximum predicted complex contained 72 proteins. With the Collins data, GECA achieved recall, precision and F-measure values of 0.498, 0.526, and 0.511, respectively when compared with the MIPS data. The other methods (ClusterONE, WCOACH, WEC, and CAG) achieved F-measure values of 0.414, 0.375, 0.499, and 0.463, respectively. The F-measure values of GECA, ClusterONE, WCOACH, WEC, and CAG were 0.438, 0.458, 0.323, 0.437, and 0.431, respectively when using SGD as benchmark data. GECA had the highest F-measure with the Collins data using MIPS as benchmark data.

Human data: GECA predicted 857 complexes, with 288 matched with the CORUM data and the maximum predicted complex contained 103 proteins. GECA achieved the recall, precision and F-measure values of 0.227, 0.336, and 0.271, respectively. ClusterONE, WCOACH, CAG, and WEC achieved F-measure values of 0.149, 0.09, 0.158, and 0.178, respectively.

It is evident that GECA had almost the highest F-measure with all data using the OS score to calculate the recall and precision values, as illustrated in Fig. 1. When Jac-Sim was used to calculate the precision and recall values, GECA was able to get the highest recall in all the data. It also got the highest F-measure except with Collins data, where it came second to ClusterONE because the precision of the latter was higher than that of GECA. It could predict only half the number of complexes predicted by GECA. GECA could predict the highest number of small and large complexes in all the data. All the results came from using Jac-Sim in the additional File S2. Thus, we can conclude that GECA predicts protein complexes better than its counterparts.

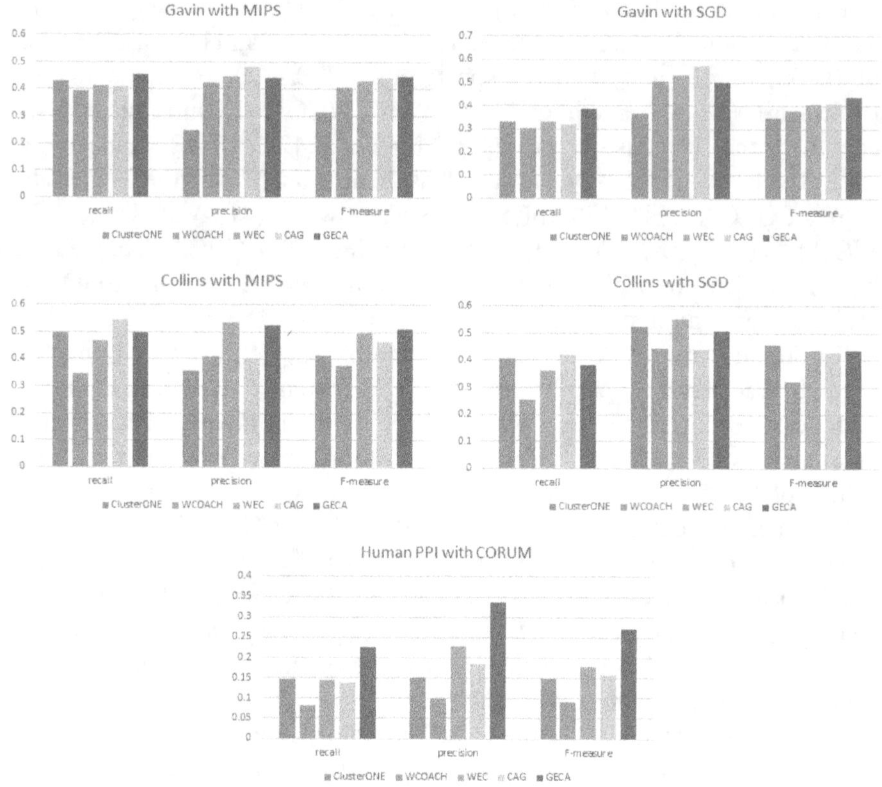

Fig. 1. Recall, precision, and F-measure of all PPI networks.

3.2. *Co-Localization and GO Semantic Similarity Scores*

Figure 2 reports the co-localization score for all methods using data from Collins and Gavin. As shown, the co-localization score of GECA is higher than those in other methods, having a value of 0.6925 using Gavin's data and 0.766 using Collins' data. An analysis of GO semantic similarity was done on the basis of the biological process (Bp), Cellular Component (CC), and Molecular Function (MF). Figure 3 shows the GO scores for all methods using Gavin, Collins, and Human PPI networks. Here, GECA achieved the highest scores for all terms using Collins and Human data, except for the Bp and CC terms which used Gavin data. The pleasing results that GECA achieved in co-localization and GO scores support the biological significance of the PCs.

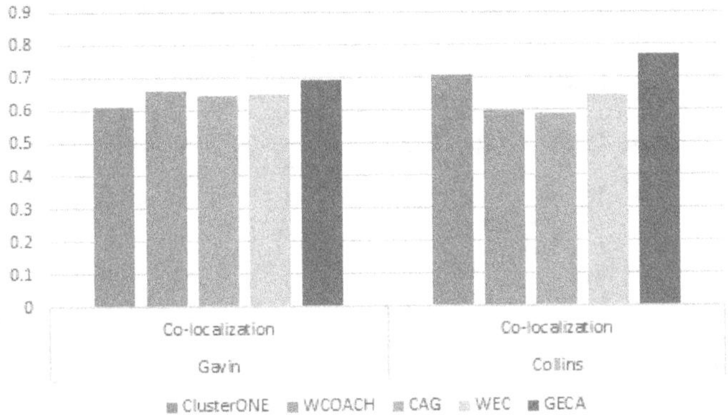

Fig. 2. Co-localization score of Gavin and Collins PPI networks.

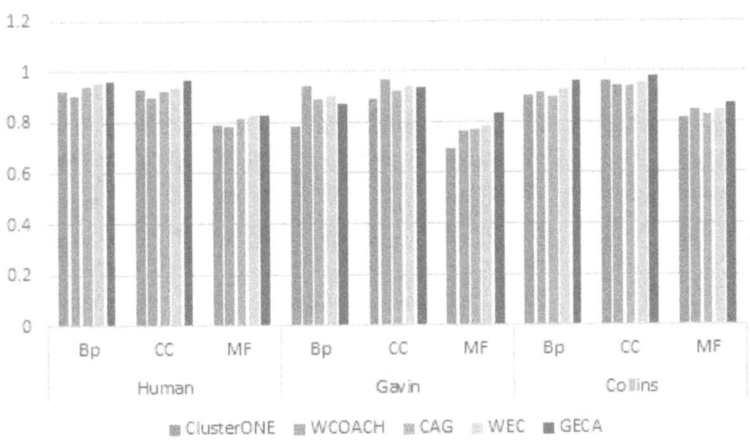

Fig. 3. GO similarity score of Gavin, Collins, and human PPI networks.

3.3. *High and Exact Matching of Actual Complexes*

In this section, the quality of PCs is evaluated by reporting the number of actual complexes that match them precisely and the number of actual complexes having an OS score equal to or greater than 0.8, but not matching exactly. GECA achieved the best number of matching complexes in all the actual data (Table 2). Hence, we can conclude that GECA predicts complexes that exactly or best match actual complexes and outperforms other algorithms in this aspect.

Table 2. High and exact matching with actual complexes.

	Gavin with MIPS		Gavin with SGD		Collins with MIPS		Collins with SGD		Human data	
	Exact	**High**	**Exact**	**High**	**Exact**	**High**	**Exact**	**High**	**Exact**	**High**
ClusterONE	7	6	14	11	14	8	33	**14**	4	6
WCOACH	**9**	9	16	**13**	4	10	3	13	2	4
WEC	8	11	16	10	14	12	36	11	6	4
CAG	6	12	12	11	15	9	33	13	4	5
GECA	8	**16**	**19**	13	**17**	**16**	**37**	**14**	**7**	**7**
Total	GECA 24		GECA 32		GECA 33		GECA 51		GECA 14	

3.4. *GSE7645 Gene Expression Results*

GECA, CAG, and WEC were implemented with the GSE7645 GEP. GSE7645 contained information about 5713 genes and covered 1537 genes of the 1622 in the Collins data and 1757 genes of the 1855 in the Gavin data. By using the OS score and Collins data with MIPS, GECA achieved recall, precision and F-measures of 0.512, 0.515, and 0.514. It was followed by WEC with 0.463, 0.54, and 0.499 and CAG with 0.517, 0.395, and 0.448. GECA predicted 15 exact complexes and 17 complexes with a high degree of matching (greater than 0.8). WEC and CAG predicted 14 exact complexes and had 11 well-predicted ones each. Gavin with CYC2008[8] was a benchmark complex. GECA achieved recall, precision and an F-measure of 0.358, 0.522, and 0.425, followed by CAG with 0.304, 0.612, and 0.406. WEC achieved 0.301, 0.567, and 0.393. GECA predicted 11 exact complexes and had 22 good predictions followed by WEC with 9 exact and 17 good predictions and CAG with 7 exact complexes and 15 good predictions. As a result, GECA was shown to outperform WEC and CAG in the F-measure and the number of well-PCs.

3.5. *Analysis of Predicted Complexes*

An important aspect of any algorithm that predicts protein complexes is that it can predict overlapping complexes since that protein can participate in multiple complexes. GECA is notably good at predicting overlapping complexes. In the Collins data with SGD, GECA predicted two overlapping complexes which exactly matched actual ones. The first

Table 3. Overlapping PCs.

PC	Actual complex	Overlapping genes
YFL049W YGR275W YBR289W YDR073W YNR023W YJL176C YMR033W YHL025W YPR034W YPL016W YOR290C	YMR033W YHL025W YFL049W YPR034W YPL129W YGR275W YPL016W YBR289W YDR073W YNR023W YOR290C YJL176C	YMR033W YPR034W YGR275W YOR290C YJL176C
YGR056W YCR020W-B YFR037C YGR275W YML127W YLR357W YJL176C YMR091C YMR033W YLR033W YDR303C YKR008W YPR034W YHR056C YIL126W YCR052W YOR290C YLR321C	YCR020W-B YGR275W YML127W YLR357W YMR033W YKR008W YPL129W YPR034W YPL082C YIL126W YCR052W YLR321C YBL006C YGR056W YFR037C YMR091C YLR033W YDR303C YHR056C YOR290C	

complex was YNL262W YBR278W YPR175W YDR121W, and the
second was YOR304W YGL133W YDR121W YJL065C, which overlapped with YDR121W. Other examples of overlap between exactly
matching predictions and actual complexes in human data were the first
complex (TCEB3C TCEB2 TCEB1) and the second (TCEB2 TCEB1
TCEB3B), which overlapped in two proteins (TCEB1 and TCEB2).
Table 3 presents a further example of overlapping complexes from the
Collins data with SGD. The two complexes overlapped with five genes
and both complexes had an OS greater than 0.8. The PCs had densities of
0.8 and 0.82.

GECA could predict complexes that contain a gene which had only a
single connection with a complex, using the gene expression information.
Figure 4(a) shows the complex that GECA predicted exactly when using
the Collins data with SGD as benchmark data. GECA was also able to
predict complexes with a density of less than 0.5. Table 4 shows the
predicted complex with a density of 0.48 and a considerable biological
significance. It had an OS of 0.35 with an actual complex in the Collins
data with SGD.

Moreover, GECA was analyzed to show the effectiveness of using GEP
and its influence on the F-measure using the OS score. Figure 4(b) shows
that the F-measure of the algorithm was the highest in all the data when

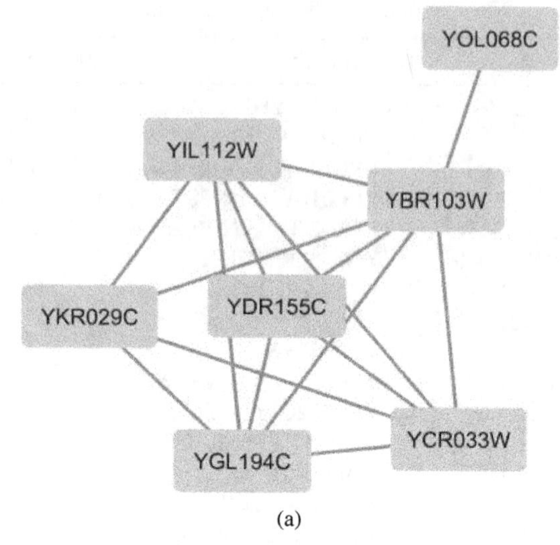

(a)

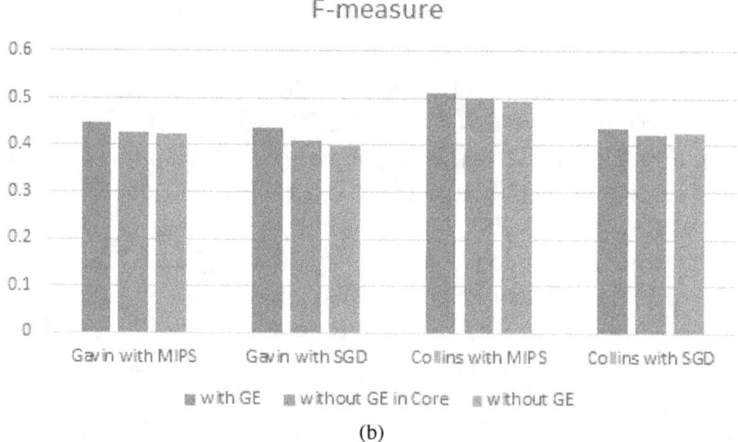

(b)

Fig. 4. (a) Exactly PC with peripheral gene. (b) F-measure of GECA with GEP and without GEP.

using GEP followed by the F-measure of the algorithm when constructing the core according to topological relations and ignoring the addition of protein to the core according to its gene expression pattern. The lowest F-measure for the algorithm is observed when constructing the core which depends on its connection with its neighbors and then attaching the closely connected gene to the core without using the GEP. As a result, applying a

Table 4. Predicted complex with density less than 0.5 and overlapping score 0.35 with actual complex.

PC	Actual complex	Co-localization	BP	CC	MF
YLR448W YPR107C YMR061W YNL317W YOR179C YBL072C YGL103W YKR002W YOR096W YPL090C YOL127W YDL082W YLL045C YGL076C YHL033C YJL190C YIL133C YFR031C-A YER133W YBR048W YNL222W YBL027W YAL043C YDL083C YKL018W YKL059C YOR063W YPL198W YGR156W YNL178W YMR142C YBR191W YJL177W YDR418W YDR301W YMR194W YLR277C YJR093C YLR441C YER074W YLR029C YDR195W YGR085C YLR115W YNL069C YPL220W	YPR107C YNL222W YLR277C YMR061W YKL018W YKL059C YJR093C YNL317W YGR156W YAL043C YOR179C YDR195W YLR115W YKR002W YER133W YDR301W	0.571	0.867	0.858	0.933

Table 5. Gene without GEP but in predicted and actual complexes.

Gene	PC	Actual complex
YHR039C-A	YDL185W **YHR039C-A** YPR036W YGR020C YEL051W YOR332W YBR127C YKL080W	YBR127C YDL185W **YHR039C-A** YPR036W YGR020C YEL051W YOR332W YOR270C YKL080W
YHR056C	YCR052W YGR056W YLR033W YMR091C YIL126W YFR037C YDR303C YCR020W-B YLR357W YMR033W YPR034W YGR275W YML127W YKR008W **YHR056C** YLR321C	YCR052W YMR091C YOR290C YCR020W-B YLR357W YML127W **YHR056C** YGR056W YLR033W YIL126W YFR037C YDR303C YBL006C YPL129W YMR033W YPR034W YGR275W YKR008W YPL082C YLR321C

gene expression in predicting protein complexes can enhance the prediction and increase its accuracy.

The most important aspect of the GECA algorithm is not merely to keep the proteins and their connections that do not have a gene expression pattern, but to determine how removing these proteins can affect the PCs. Table 5 lists some of the genes which did not have an expression pattern but were found in the PCs in the Collins data with SGD that had an OS greater than 0.8 with an actual complex.

Overall, GECA performs well, if not better than others in all cases with both yeast and human data. It also achieves the highest number of exact and well-PCs in all the data. This indicates that GECA complexes have a high matching with benchmark complexes. That is, all in all, it increases the accuracy of prediction.

4. Conclusion

GECA is a technique used to predict overlapping protein complexes from a weighted PPIN with gene expression without removing any proteins from the network as previous methods did. As such, it leads to enhanced results because it can avoid the loss of important nodes that have no gene expression information. In this study, GECA used the core-attachment technique to construct protein complexes in which the proteins of the core were either closely connected or had a similar gene expression pattern. The proteins attached to the core applied a new technique that allowed the addition of protein that was connected to fewer than half the proteins in the core but with good similarity in the gene expression pattern to proteins in the core. GECA was compared with four methods using two datasets, one concerning yeast with two GEPs and the other human. The evaluation showed that GECA performed well, if not best, regarding the F-measure, co-localization score, number of exact PCs, and GO semantic similarity scores. Moreover, the results demonstrated that GECA could effectively identify protein complexes with high biological significance.

References

1. Yong CH *et al.*, Supervised maximum-likelihood weighting of composite protein networks for complex prediction, *BMC Syst Biol* **6**(2):1–21, 2012.
2. Bader GD, Hogue CW, An automated method for finding molecular complexes in large protein interaction networks, *BMC Bioinf* **4**(1):2, 2003.
3. Altaf-Ul-Amin M *et al.*, Development and implementation of an algorithm for detection of protein complexes in large interaction networks, *BMC Bioinf* **7**(1):207, 2006.
4. Li M *et al.*, Modifying the DPClus algorithm for identifying protein complexes based on new topological structures, *BMC Bioinf* **9**(1):398, 2008.
5. Adamcsek B *et al.*, CFinder: Locating cliques and overlapping modules in biological networks, *Bioinformatics* **22**(1):1021–1023, 2006.
6. Liu G, Wong L, Chua HN, Complex discovery from weighted PPI networks, *Bioinformatics* **25**(15):1891–1897, 2009.
7. Tadaka S, Kinoshita K, NCMine: Core-peripheral based functional module detection using near-clique mining, *Bioinformatics* **32**(22):3454–3460, 2016.
8. Pu S *et al.*, Up-to-date catalogues of yeast protein complexes, *Nucleic Acids Res* **37**(3):825–831, 2009.
9. Mewes H-W *et al.*, MIPS: Analysis and annotation of proteins from whole genomes in 2005, *Nucleic Acids Res* **34**(Suppl. 1):D169–D172, 2006.
10. Aloy P *et al.*, Structure-based assembly of protein complexes in yeast, *Science* **303**(5666):2026–2029, 2004.
11. Liu G *et al.*, Decomposing PPI networks for complex discovery, *Proteome Sci* **9**(1):1–11, 2011.
12. Gavin A-C *et al.*, Proteome survey reveals modularity of the yeast cell machinery, *Nature* **440**(7084):631, 2006.
13. Leung HC *et al.*, Predicting protein complexes from PPI data: A core-attachment approach, *J Comput Biol* **16**(2):133–144, 2009.
14. Wu M *et al.*, A core-attachment based method to detect protein complexes in PPI networks, *BMC Bioinf* **10**(1):169, 2009.

15. Kouhsar M, Zare- Mirakabad F, Jamali Y, WCOACH: Protein complex prediction in weighted PPI networks, *Genes Genet Syst* **90**(5):15-00032, 2016.

16. Chin C-H *et al.*, A hub-attachment based method to detect functional modules from confidence-scored protein interactions and expression profiles, *BMC Bioinf* **11**(1):S25, 2010.

17. Keretsu S, Sarmah R, Weighted edge based clustering to identify protein complexes in protein–protein interaction networks incorporating gene expression profile, *Comput Biol Chem* **65**:69–79, 2016.

18. Keretsu S, Sarmah R, Identification of protein complexes in protein-protein interaction networks by core-attachment approach incorporating gene expression profile, *Int J Bioinf Res Applications* **13**(4):313–328, 2017.

19. Li M *et al.*, Identifying dynamic protein complexes based on gene expression profiles and PPI networks, *BioMed Res Int* **2014**:Article ID 375262, 10 pages, 2014.

20. Makrodimitris S, Reinders MJ, van Ham RC, Metric learning on expression data for gene function prediction, *Bioinformatics* **36**(4):1182–1190, 2020.

21. Radicchi F *et al.*, Defining and identifying communities in networks, *Proc Natl Acad Sci* **101**(9):2658–2663, 2004.

22. Dezső Z, Oltvai ZN, Barabási A-L, Bioinformatics analysis of experimentally determined protein complexes in the yeast Saccharomyces cerevisiae, *Genome Res* **13**(11):2450–2454, 2003.

23. Li X-L, Foo C-S, Ng S-K, Discovering protein complexes in dense reliable neighborhoods of protein interaction networks, *Comput Syst Bioinf* **6**:157–168, 2007.

24. Yong CH, Wong L, From the static interactome to dynamic protein complexes: Three challenges, *J. Bioinf Comput Biol* **13**(2):1571001, 2015.

25. Jansen R, Gerstein M, Analyzing protein function on a genomic scale: The importance of gold-standard positives and negatives for network prediction, *Curr Opin Microbiol* **7**(5):535–545, 2004.

26. Jansen R *et al.*, A Bayesian networks approach for predicting protein-protein interactions from genomic data, *Science* **302**(5644):449–453, 2003.

27. Wang JZ *et al.*, A new method to measure the semantic similarity of GO terms, *Bioinformatics* **23**(10):1274–1281, 2007.

28. Schlicker A *et al.*, A new measure for functional similarity of gene products based on Gene Ontology, *BMC Bioinf* **7**(1):302, 2006.

29. Collins SR *et al.*, Toward a comprehensive atlas of the physical interactome of Saccharomyces cerevisiae, *Mol Cell Proteomics* **6**(3):439–450, 2007.

30. Tu BP *et al.*, Logic of the yeast metabolic cycle: Temporal compartmentalization of cellular processes, *Science* **310**(5751):1152–1158, 2005.

31. Mewes H-W *et al.*, MIPS: Analysis and annotation of proteins from whole genomes, *Nucleic Acids Res* **32**(Suppl. 1):D41–D44, 2004.

32. Dwight SS *et al.*, Saccharomyces Genome Database (SGD) provides secondary gene annotation using the Gene Ontology (GO), *Nucleic Acids Res* **30**(1):69–72, 2002.
33. Peri S *et al.*, Development of human protein reference database as an initial platform for approaching systems biology in humans, *Genome Res* **13** (10):2363–2371, 2003.
34. Liu Q, Song J, Li J, Using contrast patterns between true complexes and random subgraphs in ppi networks to predict unknown protein complexes, *Sci Rep* **6**:21223, 2016.
35. Nymark P *et al.*, Gene expression profiles in asbestos-exposed epithelial and mesothelial lung cell lines, *BMC Genom* **8**(1):62, 2007.
36. Ruepp A *et al.*, CORUM: The comprehensive resource of mammalian protein complexes — 2009, *Nucleic Acids Res* **38**(Suppl. 1): D497–D501, 2010.
37. Nepusz T, Yu H, Paccanaro A, Detecting overlapping protein complexes in protein-protein interaction networks, *Nat Methods* **9**(5):471, 2012.

Soheir Noori received B.Sc. degree in Computer Science from the University of Babylon, Iraq, in 2001, and M.Sc. degree in Computer Science from the University Putra Malaysia, Malaysia in 2014. After completing her M.Sc., she worked as Assistant Lecturer at the Department of Computer Science, the University of Kerbala. In 2018, she began her Ph.D. in Computer Science at the University of Babylon. Her research interests include data structure, bioinformatics, human-computer interaction, and data mining.

Nabeel H Al-A'araji received B.Sc. in Mathematics from the Al-Mustansiryah University, Iraq, in 1976, M.Sc. in Mathematics from the University of Baghdad, Iraq in 1978 and Ph.D. in Mathematics from the University of Wales, Aberystwyth, UK, in 1988. He is a Professor of Computer Science since 1997. He was Head of the Department of Computer Science for 12 years, and served as the Chancellor of the University of Babylon for 8 years and Advisor of higher Education Minister for one year. Currently, he is the Head of Supervision and Scientific Evaluation Apparatus in the Higher Education Ministry, Iraq. His research interests include artificial intelligence, GIS machine learning, neural networks, deep learning, and data mining.

Eman Al-Shamery received B.Sc. and M.Sc. degrees in Computer Science from the University of Babylon, Iraq, in 1998 and 2001, respectively. After completing her M.Sc., she worked as Assistant Lecturer at the Department of Computer Science, the University of Babylon. In 2013, she received her Ph.D. in Computer Science from the University of Babylon. Currently, she is Professor at the Department of Software, the University of Babylon. Her research interests include artificial intelligence, bioinformatics, machine learning, neural networks, deep learning, and data mining.

CHAPTER 4

Integrating Network Topology, Gene Expression Data and GO Annotation Information for Protein Complex Prediction[a]

Wei Zhang[*,¶], Jia Xu[†], Yuanyuan Li[‡] and Xiufen Zou[§]

*School of Science, East China Jiaotong University
Nanchang 330013, P. R. China*

*†School of Mechatronic Engineering
East China Jiaotong University
Nanchang 330013, P. R. China*

*‡School of Mathematics and Statistics
Wuhan Institute of Technology in Wuhan
Wuhan 430072, P. R. China*

*§School of Mathematics and Statistics
Wuhan University, Wuhan 430072, P. R. China*
¶wzhang_math@whu.edu.cn

The prediction of protein complexes based on the protein interaction network is a fundamental task for the understanding of cellular life as well as the mechanisms underlying complex disease. A great number of methods have been developed to predict protein complexes based on protein–protein interaction (PPI) networks in recent years. However, because the high throughput data obtained from experimental biotechnology are incomplete, and usually contain a large number of spurious interactions, most of the network-based protein complex identification methods are sensitive to the reliability of the PPI network. In this paper,

---

[a]This article was previously published in *Journal of Bioinformatics and Computational Biology*. Vol: 17, No. 1 (2019), 1950001 (20 pages).

we propose a new method, Identification of Protein Complex based on Refined Protein Interaction Network (IPC-RPIN), which integrates the topology, gene expression profiles and GO functional annotation information to predict protein complexes from the reconstructed networks. To demonstrate the performance of the IPC-RPIN method, we evaluated the IPC-RPIN on three PPI networks of *Saccharomycescerevisiae* and compared it with four state-of-the-art methods. The simulation results show that the IPC-RPIN achieved a better result than the other methods on most of the measurements and is able to discover small protein complexes which have traditionally been neglected.

Keywords: PPI network; protein complexes; GO ontology; gene expression profile.

1. Introduction

Protein complexes are formed by groups of physically interacting proteins into functional units that are integral to maintaining normal physiological activities.[1] Discovering protein complexes is one of the key areas of research for understanding the cellular function and the dynamic mechanisms underlying complex disease. Experimentally, the identification of protein complexes is typically performed by the use of Tandem Affinity Purification with mass spectrometry (TAP-MS).[2] However, these biological experiments are expensive and time-consuming.

Due to the rapid development of modern high-throughput technologies such as yeast two-hybrid (Y2H) screens,[3,4] tandem affinity purification (TAP),[5] and mass spectrometric protein complex identification (MS-PCI),[6] large scale protein–protein interaction (PPI) data are available for many organisms. PPI networks provide a comprehensive view of the global interaction structure of an organism's proteome, as well as detailed information on specific interactions, which make it possible and feasible to use computational methods to predict protein complexes from the perspective of large molecular network.

The past decade has witnessed the rapid development of computational methods for identifying protein complexes from PPI network datasets. Many approaches have been developed to predict protein complexes,[7–13,18–25,27,28] such as MCL,[7] MCODE,[8] COACH,[11] CMC,[12] ClusterOne,[13] PEWCC,[18] WPNCA,[19] and WCOACH.[22] These methods basically fall into three categories: seed expanding-based methods, clustering-based methods and dynamic PPI-based methods. The first category identifies protein complexes based on seed expanding or core expanding by capturing the density of subgraphs in PPI networks.[8,11–13] For example, Liu *et al.*[12] developed an iterative scoring method (AdjstCD) to evaluate the reliability of interaction between protein pairs by assigning weight to protein pairs and predicting protein complexes in weighted PPI networks. Wu *et al.*[11] proposed a core-attachment based method, COACH, which first identifies the core of protein complexes and then incorporates the attachments associated with the core proteins. Nepusz *et al.*[13] proposed a seed growth method for predicting protein complexes, named ClusterOne, by defining a cohesiveness score for a subnetwork.

It has been widely accepted that the efficient and effective integration of multiple data sources yields better results.[29,30] Han Yong *et al.*[31] proposed a supervised maximum-likelihood method to weight edges in a composite network constructed by heterogeneous data and adapted clustering algorithms for protein complex prediction. Wu *et al.*[32] developed an integrative approach, named InteHC, which utilized the Support Vector Machine (SVM) to weight features for heterogeneous data and generated protein complexes by using the hierarchical clustering algorithm. Comparative analysis showed that InteHC outperformed the other 14 methods. Recently, methods have incorporated additional information including Gene Ontology (GO),[33,34] gene expression data[35] and structural interface data of protein domains[40] to help predict protein complexes. The simulation results of these methods show that integrating multiple sources of data can boost the detection of protein complexes.

The second category consists of clustering the proteins based on clustering and evolutional algorithms.[14–17,36] For example, Ou-Yang *et al.*[36] proposed a new protein prediction method based on nonnegative matrix factorization. Sanghamitra *et al.*[15] proposed a multi-objective framework for predicting protein complexes based on both topological properties of PPI network and GO semantic similarity information.

The methods in the third category predict protein complexes by constructing dynamic PPI networks. Zhang *et al.*[37] identify protein complexes by constructing dynamic probabilistic protein networks. Ou-Yang *et al.*[38] developed a novel time smooth overlapping complex detection model (TS-OCD) for predicting temporal protein complexes and tracking the evolutionary process of the temporal complexes. Li *et al.*[39]

proposed a model-based scheme for construction of the spatial and temporal protein interaction network by combing gene expression and subcellular location information, and the algorithm MCL is used for predicting protein complexes based on the temporal network.

However, most of the methods described above only focus on the original network obtained from high-throughput techniques, and few of them consider the inherent incompleteness and noise of the PPI networks.[41]

Due to the fact that PPI data are inherently noisy and incomplete, the incompleteness and noise underlying the PPI networks severely hindered the prediction accuracy of these methods.

In this research, we try to introduce a novel method (IPC-RPIN) for predicting protein complexes by constructing a refined PPI network based on integrating gene expression data and GO ontology information. To validate the performance of the IPC-RPIN, we compare it with four other state-of-the-art methods (ClusterOne, PEWCC, COACH, and TNC (Ref. 24)) on three benchmark yeast PPI datasets. The simulation results show that the new method performs well in identifying protein complexes.

2. Method

2.1. *Protein Complex Prediction from Refined Network*

In this section, we first present a new strategy for obtaining a refined PPI network by link prediction based on gene expression profiles, and then we introduce the new proposed method based on the definition of protein complexes.

Given a PPI network with N proteins, usually represented by an undirected graph $G = (V, E)$, the vertex set V represents the proteins and the edge set E represents the set of interactions between pairs of proteins.

Previous works have demonstrated that integrating multiple data sources could improve the prediction accuracy. In addition, constructing new networks by PPI prediction based on gene expression profiles demonstrates more reliable for mining key information such as essential proteins.[42,43] So, we first constructed a new PPI network by PPI prediction based on gene expression profiles and then incorporated GO annotation information to qualify the compactness of the subnetwork.

The new proposed algorithm is as follows:

Algorithm 1. Prediction protein complexes

Input: The PPI network $G = (V, E)$; GE: gene expression data; GO:Gene ontology annotation data.

Output: Predicted protein complexes (PPC)

1: **for** each unlinked protein pairs $(u, v) \notin E$ **do** Compute the PCC(u, v) between two proteins u and v.

2: **if** PCC$(u, v) \geq 0.98$ **then** Add the link (u, v) to the original network G.

3: Obtain new network G'

4: Compute the GO semantic similarity matrix G simgo based on GO annotation information under Biological Process term.

5: **for** each protein u $\in$ V **do**

6: Create a rough group Vg by adding the neighbors of u.

7: Evaluate the compactness of the rough group $CVgt$(Topology compactness of rough group), $CVgo$ (GO compactness of rough group).

8: Add vertices on the outer boundary of the Vg iteratively to increase the compactness $CVgt$ and $CVgo$ until they reaches the maximum.

9: Remove vertices in the Vg if the compactness of the remaining group $CVgt$ and $CVgo$ are larger than original, and iteratively increase the compactness until it reaches the maximum.

10: Obtain initial protein complex set PCS

11: Merge the pairs of candidate connected protein complexes with OS ≥ 0.8 and obtain refined protein complexes sets.

12: **for** each protein complex pc $\in$ PCS **do**

13: **if** size(pc)<3 or density(pc)<0.5 **then**

14: delete the pc from PCS

15: Obtain the last predicted protein complex PPC

16: **return** Predicted protein complexes PPC

The topology compactness of a group $(CVgt)$ is adapted from our previous work[24]

$$CVgt(G') = \frac{Nl^{\text{in}}(G')}{Nl^{\text{in}}(G') + Nl^{\text{out}}(G') + p|G'|}, \tag{1}$$

where the $Nl^{\text{in}}(G')$ denotes the total number of 3-cliques in the subgraph G', and the $Nl^{\text{out}}(G')$ denotes the total number of 3-cliques that connect the subgraph with the rest of the network, $p|G'|$ is a penalty term which measures the inaccuracy of the network interaction. The value of p is set to

0.1 in this work and the effect of parameter p on the results is discussed in Sec. 3.4.

The new compactness after the addition of protein i (protein i is a member of Vb') can be calculated as follows:

$$CVgt(G' \cup \{i\}) = \frac{Nl^{\text{in}}(G') + Nl^{\text{in}}(i)}{Nl^{\text{in}}(G') + Nl^{\text{out}}(G') + Nl^{\text{out}}(i) + p(|G'| + 1)}, \quad (2)$$

where the $Nl^{\text{in}}(i)$ denotes the total number of 3-cliques that connect protein i with proteins in subgraph G', and $Nl^{\text{out}}(i)$ denotes the total number of 3-cliques that connect protein i with nonmembers of G'.

Similarly, the new compactness after removal of protein i (in subgraph G') can be calculated as follows:

$$CVgt(G' \setminus \{i\}) = \frac{Nl^{\text{in}}(G') - Nl^{\text{in}}(i)}{Nl^{\text{in}}(G') + Nl^{\text{out}}(G') - Nl^{\text{out}}(i) + p(|G'| - 1)}. \quad (3)$$

The GO compactness of rough group $G'(V', E')$ is defined as

$$CVgo(G') = \frac{\sum_{u,v \in E'} \text{GO\_sim}(X, Y)}{|G'|(|G'| - 1)/2}. \quad (4)$$

The $\text{GO\_sim}(X, Y)$ is the GO functional similarity between two proteins a and b, which is calculated by using the GO semantic similarity. To evaluate the gene functional similarity between GO terms annotated to proteins in the interaction networks, the method developed by Wang et al.[44] is applied to calculate the semantic similarity between proteins. In this work, we only consider the BP subontology term in evaluating the two considered proteins. The GO similarity between two connected proteins is defined as

$$\text{GO\_sim}(X, Y) = \frac{\sum_{1 \le i \le m} s(go_{x_i}, Y) + \sum_{1 \le j \le n} s(go_{y_j}, X)}{m + n}, \quad (5)$$

where $s(go_{x_i}, Y) = \max_{1 \le j \le n}(S\_GO(go_{x_i}, go_{y_j}))$, $s(go_{y_j}, X) = \max_{1 \le i \le n}(S\_GO(go_{x_i}, go_{y_j}))$, and $S\_GO(go_{x_i}, go_{y_j})$ is the semantic similarity between terms go_{x_i} and go_{y_j}. The semantic similarity between two considered terms A and B is defined as

$$S\_GO(A, B) = \frac{\sum_{r \in T_A \cap T_B}(S_A(r) + S_B(r))}{\sum_{r \in T_A} S_A(r) + \sum_{r \in T_B} S_B(r)}, \quad (6)$$

where $S_A(r)$ is the S-value of GO term r related to term A and $S_B(r)$ is the S-value of GO term r related to term B.

2.2. *Experimental Data*

Genome-wide PPI networks are publicly available in several open databases. The yeast *Saccharomycescerevisiae* PPI networks are widely used as benchmarks for evaluating the performance of a newly proposed algorithm, as they have been well characterized by biological experiments.

In our proposed method, we used three different yeast PPI datasets including the collins2007 dataset, which contains 9074 interactions and 1622 proteins, the gavin2006 dataset, which contains 1855 proteins and 7669 interactions, and the krogan_extended dataset, which consists 14317 interactions and 3672 proteins. The three datasets are obtained from the published work in Ref. 13. In the following statement, we use collins, gavin and krogan_ extended to represent these three networks.

Three benchmark reference complexes are used to evaluate the performance of the new method: the CYC2008 dataset,[45] which contains 408 manually curated heteromeric protein complexes, the Mips dataset,[46] which contains 203 protein complexes and the Aloy dataset,[47] which contains 101 protein complexes.

The gene expression data used in our experiment are from GSE3431 dataset,[48] which collects the data of 12 time points during three successive metabolic cycles and approximately 25 min per time interval.

2.3. *Evaluation Measures*

To evaluate the efficiency of the proposed method, we compare the IPC-RPIN algorithm to the following state-of-the-art algorithms: ClusterOne, PEWCC and TNC. All methods are performed on the refined networks. There are four different quantity measures, namely, the number of matched protein complexes (Match) in the reference protein complexes, the fraction of reference protein complexes predicted (Frac), the prediction geometric accuracy (ACC),[49] and maximum matching ratio (MMR).[13]

The Overlapping Score (OS)[8] between a predicted protein complex P and a known protein complex R is defined as follows:

$$OS(R, P) = \frac{|R \cap P|^2}{|R||P|},$$ (7)

where R and P represent the benchmark reference protein complexes and predicted protein complexes, respectively. We assume the reference complex has been predicted when the predicted protein complex and reference complexes with an OS ≥ 0.25.

The MMR is based on a maximal one-to-one mapping between predicted and reference complexes.

$$\text{MMR}(R, P) = \frac{\sum_{i=1}^{n} \max_{j=1}^{m} \text{OS}(R_i, P_j)}{n},\tag{8}$$

where the OS refers to the overlap score between two protein sets.

The Acc, is the geometric mean of the clustering-wise sensitivity (Sn) and the clustering-wise positive predictive value (PPV). Given m predicted and n reference protein complexes, the two measures are based on the confusion matrix $T = [t_{ij}]$ of the complexes, where t_{ij} denotes the number of proteins that are found both in reference complex i and predicted complex j. The Sn and PPV are defined as

$$S_n = \frac{\sum_{i=1}^{n} \max_{j=1}^{m} t_{ij}}{\sum_{i=1}^{n} n_i},\tag{9}$$

where n_i is the number of proteins in reference complex i.

$$\text{PPV} = \frac{\sum_{j=1}^{m} \max_{i=1}^{n} t_{ij}}{\sum_{j=1}^{m} \sum_{i=1}^{n} t_{ij}}.\tag{10}$$

The geometry accuracy (Acc) represents a tradeoff between sensitivity and positive predictive value and is defined as:

$$\text{Acc} = \sqrt{S_n * PPV}.\tag{11}$$

The high geometric accuracy (Acc) indicates that the two criteria of the Sn and the PPV metric are indicative of high performance.

3. Results

In this section, we first systematically evaluate the performance of the new method and against four other existing methods (ClusterOne, PEWCC, TNC, and COACH) on the three test PPI networks using four evaluation measurements. Next, we compare the evaluation results under different combination strategies in Sec. 3.3. Last, the effects of the parameters on the evaluation results are analyzed. The parameters in ClusterOne,

PEWCC and TNC are set as default. The minimum number of proteins in protein complexes is 3, the overlap score is set to 0.25, and the density threshold of the predicted protein complexes is set to 0.5 as default for the new method.

3.1. *Comparison with the Benchmark Protein Complexes*

Table 1 shows the comparative results of the new method and four existing methods on the three datasets using four evaluating measurements under the CYC2008 reference protein complex dataset. We can see that the new proposed method IPC-RPIN performs better than the other considered methods in most of these measurements. Especially, for the Collins dataset, the predicted protein complexes could match 112 reference protein complexes, which is better than other methods.

Similarly, we applied the methods on the three test datasets with respect to two other reference complex datasets. The performance comparison of the three datasets under the Mips reference complexes and Aloy reference complexes are presented in Tables 2 and 3, respectively.

As shown in Table 2, in addition to the ACC measurement, the proposed method shows superiority to other methods in the other three

Table 1. Result of the three methods on the original network and new constructed network under CYC2008 reference protein complex dataset.

PPI data	Method	Matched	Frac	Acc	MMR
Collins	IPC-RPIN	**112**	**0.572**	0.599	**0.416**
	TNC	97	0.538	0.608	0.388
	ClusterOne	96	0.504	**0.640**	0.377
	PEWCC	91	0.492	0.591	0.365
	COACH	98	0.513	0.607	0.383
Krogan_extended	IPC-RPIN	**96**	**0.551**	0.584	**0.356**
	TNC	88	0.496	**0.595**	0.331
	ClusterOne	91	0.487	0.563	0.342
	PEWCC	81	0.462	0.518	0.308
	COACH	93	0.496	0.530	0.338
Gavin	IPC-RPIN	**95**	**0.521**	0.576	**0.355**
	TNC	84	0.521	**0.598**	0.344
	ClusterOne	69	0.381	0.502	0.262
	PEWCC	88	0.492	0.590	0.340
	COACH	92	0.504	0.591	0.346

Table 2. Result of the three methods on the original network and new constructed network under Mips reference protein complex dataset.

PPI data	Method	Matched	Frac	Acc	MMR
Collins	IPC-RPIN	**74**	**0.537**	0.448	**0.360**
	TNC	66	0.512	**0.456**	0.341
	ClusterOne	67	0.512	0.397	0.332
	PEWCC	61	0.468	0.394	0.322
	COACH	66	0.507	0.437	0.339
Krogan_extended	IPC-RPIN	**58**	**0.443**	0.335	**0.288**
	TNC	53	0.399	**0.338**	0.265
	ClusterOne	54	0.404	0.336	0.266
	PEWCC	53	0.414	0.312	0.256
	COACH	53	0.429	0.318	0.271
Gavin	IPC-RPIN	**67**	**0.488**	0.368	**0.321**
	TNC	58	0.458	0.366	0.303
	ClusterOne	45	0.374	0.287	0.243
	PEWCC	62	0.468	0.374	0.317
	COACH	62	0.458	**0.375**	0.315

Table 3. Result of the three methods on the original network and new constructed network under Aloy reference protein complex dataset.

PPI data	Method	Matched	Frac	Acc	MMR
Collins	IPC-RPIN	**68**	**0.936**	0.793	**0.743**
	TNC	65	**0.936**	0.804	0.719
	ClusterOne	64	0.897	**0.842**	0.702
	PEWCC	60	0.846	0.800	0.654
	COACH	66	0.910	0.827	0.729
Krogan_extended	IPC-RPIN	**50**	**0.782**	0.764	0.500
	TNC	48	0.718	0.769	0.475
	ClusterOne	34	0.590	0.689	0.383
	PEWCC	44	0.679	0.714	0.467
	COACH	49	0.756	0.729	**0.507**
Gavin	IPC-RPIN	**67**	**0.974**	0.826	**0.685**
	TNC	65	**0.974**	**0.847**	0.684
	ClusterOne	53	0.756	0.720	0.550
	PEWCC	64	0.897	0.831	0.667
	COACH	**67**	0.949	0.841	0.682

measurements. Our method always achieves the highest Match number and MMR score.

Similar results were obtained using the Aloy reference protein complex dataset, as seen from Table 3, where the new method performs better than the other methods in most of the four measurements under all three considered datasets, suggesting that the new method is capable of detecting more true complexes and provides a strong argument in favor of the proposed method.

3.2. *Protein complexes identified by our method*

To further exhibit the performance of the new proposed method, we collected the matched protein complexes that could only be detected by the new method under the three PPI networks with respect to the three reference protein complex datasets.

The protein complexes that could only be predicted by the new method under the collins PPI network are listed in Fig. 1. Complexes A, B, D and E in Fig. 1 are evaluated under CYC2008 reference protein complexes. The three small protein complexes, A, B and C were identified accurately by the new method using the Aloy reference protein complexes dataset.

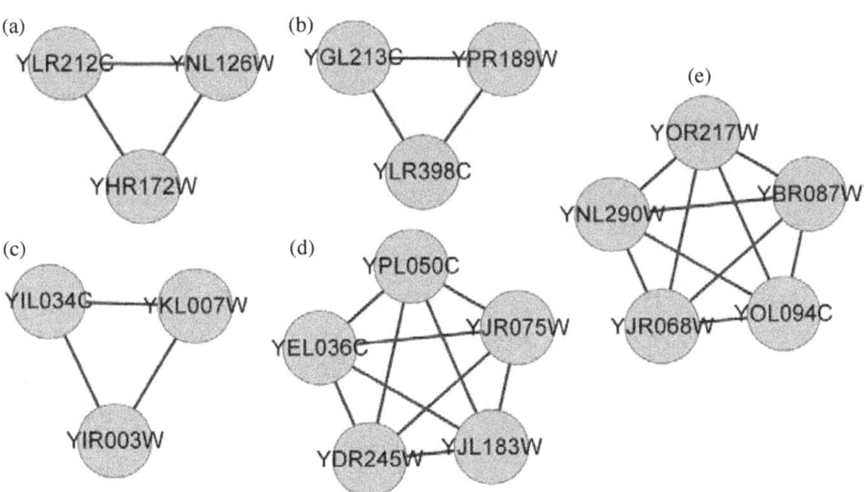

Fig. 1. The protein complexes could only be detected by using our method under the collins PPI dataset with respect to the CYC2008, Aloy and Mips reference protein complexes datasets.

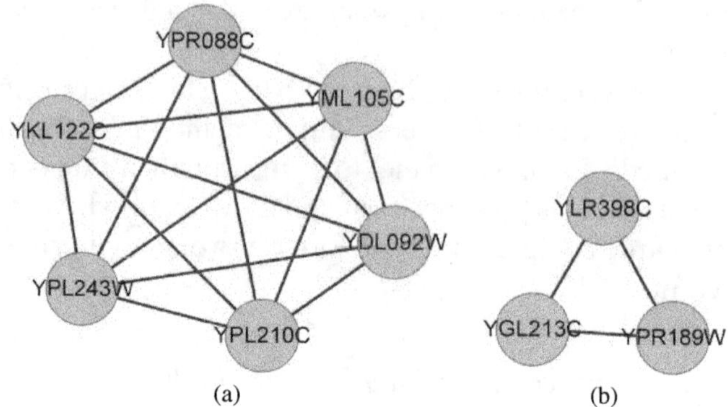

Fig. 2. The protein complexes could only be detected by using our method under the Gavin PPI dataset with respect to the CYC2008, Aloy and Mips reference protein complexes datasets.

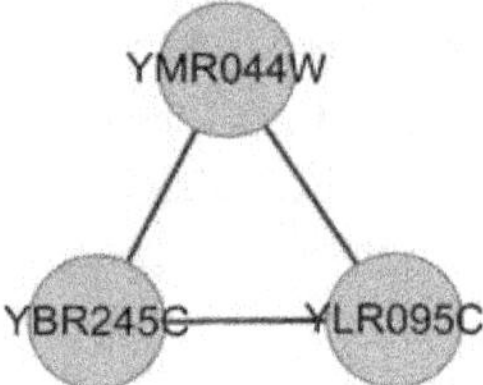

Fig. 3. The protein complexes could only be detected by using our method under the Krogan_extended PPI dataset with respect to the CYC2008, Aloy and Mips reference protein complexes datasets.

Complexes A and E predicted by the new method were matched by using the Mips reference protein complex dataset.

Figure 2 shows the protein complexes identified by the new method for the Gavin PPI dataset. Both complexes A and B could only be detected when evaluated by using the CYC2008 and Aloy reference protein complex datasets, however, complex A could only be detected by the new method for the Gavin dataset under the Mips reference protein complex dataset.

The protein complexes that could only be predicted by the new method for the Krogan_extended PPI network under the CYC2008 reference protein complex dataset are shown in Fig. 3. The small protein complexes that were accurately identified by using the new algorithm indicates that constructing refined network and adding GO information when measuring compactness of subnetworks will help in the detection of small protein complexes that were ignored by all four state-of-the-art methods.

3.3. *Comparing results under different types of strategies and threshold parameters*

The new proposed algorithm involved three types of biological data, and different strategies were used to evaluate the cohesiveness of protein complexes. It was necessary to analyze the performance under different combination of data and strategies. To test the performance of the new proposed strategy under different scenarios of additional data combination, we compared the results with different PCC thresholds under which both topological information and GO information are considered, only topological information is considered, and only GO information is considered in measuring the compactness of subnetwork. For simplicity, we set the threshold to 0.98 and 0.96, respectively.

As shown in Table 4, we list the evaluation results under different scenarios of information used in measuring the compactness of the subnetwork with different PCC thresholds. The only TNC in the bracket of the method column means that we only consider the topological compactness when performing the simulation of the IPC-RPIN algorithm, and only GO denotes that only GO compactness is calculated in step 7 of

Table 4. Evaluation results under different threshold or strategy on corresponding PPI networks with respect to the CYC2008 reference protein complexes dataset.

PPI data	Method	PCC threshold	Matched	Frac	Acc	MMR
Collins	IPC-RPIN (only TNC)	PCC = 0.98	97	0.534	0.595	0.384
	IPC-RPIN (only GO)		70	0.407	0.502	0.262
	IPC-RPIN (only TNC)	PCC = 0.96	95	0.521	0.555	0.378
	IPC-RPIN (only GO)		70	0.407	0.504	0.262
Krogan_extended	IPC-RPIN (only TNC)	PCC = 0.98	81	0.479	0.575	0.318
	IPC-RPIN (only GO)		74	0.386	0.462	0.268
	IPC-RPIN (only TNC)	PCC = 0.96	76	0.462	0.479	0.303
	IPC-RPIN (only GO)		73	0.386	0.461	0.264
Gavin	IPC-RPIN (only TNC)	PCC = 0.98	82	0.508	0.584	0.332
	IPC-RPIN (only GO)		72	0.419	0.486	0.263
	IPC-RPIN (only TNC)	PCC = 0.96	80	0.492	0.489	0.324
	IPC-RPIN (only GO)		71	0.407	0.484	0.258

the algorithm. Comparing Table 4 with results listed in Table 1 for IPC-RPIN, performance is slightly worse when only TNC is considered for three PPI networks under the CYC2008 reference protein complex dataset. When only the GO compactness is considered in the algorithm, the results are the worst. The comparison results indicated that the appropriate integration of different data for evaluation compactness will increase the performance of protein complex prediction and that the topological information is more useful than GO information when only one dataset is considered.

The evaluation results of the Mips and Aloy benchmark reference protein complex datasets are shown in Tables 5 and 6, respectively. Similar results can be seen from the two tables, in which IPC-RPIN algorithm performs better than the original TNC-based method and when only one type of data information is considered in measuring the compactness of the subgraph. These results suggest that the addition of GO annotation information could improve the performance of predicting protein complexes.

In addition, the results under different PCC thresholds show that when the PCC is set to be relatively large (PCC = 0.98), the performance of the four considered measurements is better than when the PCC is set to be relatively small (PCC = 0.96), indicating that when the PCC threshold is set to be relatively large, the added edges may include the real missed

Table 5. Evaluation results under different threshold or strategy on corresponding PPI networks with respect to the Mips reference protein complexes dataset.

PPI data	Method	PCC threshold	Matched	Frac	Acc	MMR
Collins	IPC-RPIN (only TNC)	PCC = 0.98	66	0.512	0.445	0.339
	IPC-RPIN (only GO)		46	0.374	0.346	0.225
	IPC-RPIN (only TNC)	PCC = 0.96	64	0.507	0.421	0.338
	IPC-RPIN (only GO)		46	0.374	0.347	0.225
Krogan_extended	IPC-RPIN (only TNC)	PCC = 0.98	55	0.404	0.334	0.268
	IPC-RPIN (only GO)		48	0.360	0.289	0.224
	IPC-RPIN (only TNC)	PCC = 0.96	48	0.409	0.306	0.263
	IPC-RPIN (only GO)		48	0.369	0.295	0.223
Gavin	IPC-RPIN (only TNC)	PCC = 0.98	58	0.443	0.363	0.294
	IPC-RPIN (only GO)		47	0.394	0.308	0.240
	IPC-RPIN (only TNC)	PCC = 0.96	56	0.448	0.332	0.294
	IPC-RPIN (only GO)		47	0.384	0.307	0.236

Table 6. Evaluation results under different threshold or strategy on corresponding PPI networks with respect to the Aloy reference protein complexes dataset.

PPI data	Method	PCC threshold	Matched	Frac	Acc	MMR
Collins	IPC-RPIN (only TNC)	PCC = 0.98	65	0.936	0.778	0.711
	IPC-RPIN (only GO)		46	0.705	0.672	0.452
	IPC-RPIN (only TNC)	PCC = 0.96	62	0.897	0.694	0.692
	IPC-RPIN (only GO)		47	0.705	0.666	0.458
Krogan_extended	IPC-RPIN (only TNC)	PCC = 0.98	47	0.731	0.758	0.479
	IPC-RPIN (only GO)		42	0.615	0.640	0.421
	IPC-RPIN (only TNC)	PCC = 0.96	42	0.654	0.654	0.445
	IPC-RPIN (only GO)		42	0.615	0.636	0.415
Gavin	IPC-RPIN (only TNC)	PCC = 0.98	65	0.962	0.836	0.665
	IPC-RPIN (only GO)		51	0.821	0.702	0.512
	IPC-RPIN (only TNC)	PCC = 0.96	63	0.923	0.708	0.643
	IPC-RPIN (only GO)		50	0.795	0.692	0.497

Table 7. Performance of new method on three PPI datasets with the density threshold of 0.3.

Reference date	PPI date	Matched	Frac	Acc	MMR
Mips	Collins	**77**	**0.557**	**0.449**	**0.373**
	Krogan_extended	64	0.498	0.339	0.318
	Gavin	68	0.502	0.371	0.328
CYC2008	Collins	**121**	**0.614**	**0.597**	**0.447**
	Krogan_extended	101	0.610	0.569	0.387
	Gavin	97	0.525	0.552	0.360
Aloy	Collins	**69**	**0.974**	0.777	**0.761**
	Krogan_extended	51	0.821	0.754	0.525
	Gavin	67	**0.974**	**0.801**	0.685

interactions between proteins and the new obtained network will be more reliable and complete for protein complex detection.

In this experiment, we evaluate the effect of the density threshold parameters of our method on the considered three PPI networks. Table 7 shows the performance of the new method under the three benchmark reference protein complexes; when the density threshold in the last step of IPC-PRIN is set to 0.3, the Match number of predicted protein complexes and fractions and MMR measurements are higher than the results obtained by setting the density to 0.5 as default. Especially for the Collins PPI network, the number of matched protein complexes under the density threshold of 0.3 is 121, which is larger than the number of matched protein complexes (112) under the density threshold of 0.5. This finding indicated that the benchmark protein complexes are sparse; however, when the density threshold is set to be relatively large, some protein complexes with sparse density may be ignored. Almost the same results can be obtained by setting the density threshold to 0, indicating that the density of the predicted protein complexes is equal or larger than 0.3.

3.4. *Influence of parameter p*

The penalty term p in the formulas of measuring the compactness of protein complexes is used to fine-tune the inaccuracy of the network. Hence, it is necessary to analyze the effects of parameter p on the performance of the new method. Figures 4–6 show the three core measurements as a function of parameter p when parameter p is varied from 0 to 0.3. We can observe that parameter p has slight effect on the core evaluation measurements under different PPI networks and benchmark reference protein complex datasets, and when parameter p varies in the

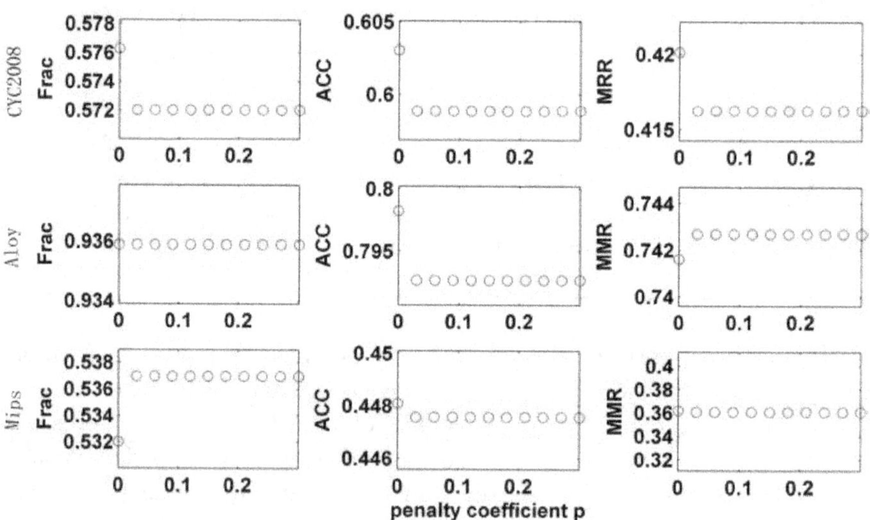

Fig. 4. The three core measurement plots as a function of penalty term p under the Collins PPI network. The three rows of the figures (on the left, middle and right) are corresponding to the CYC2008, Aloy and Mips reference protein complexes datasets.

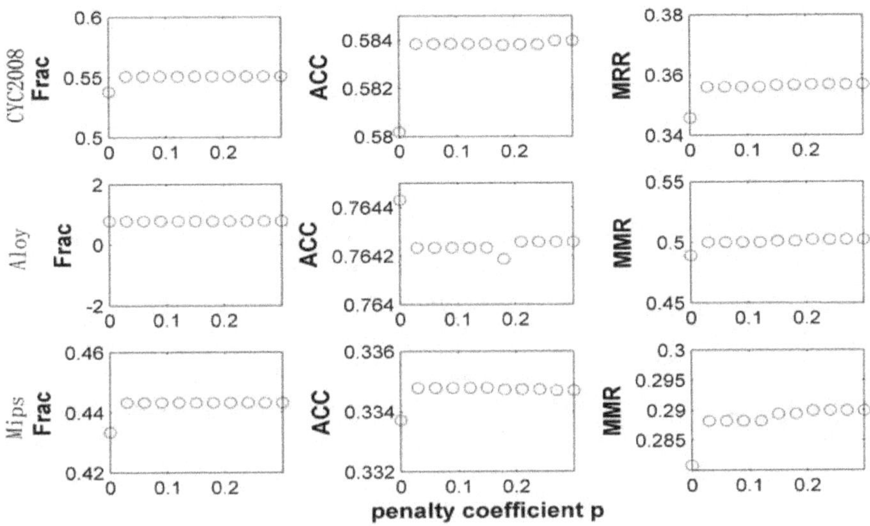

Fig. 5. The three core measurement plots as a function of penalty term p under the Krogan_extended PPI network. The three rows of the figures (on the left, middle and right) are corresponding to the CYC2008, Aloy and Mips reference protein complexes datasets.

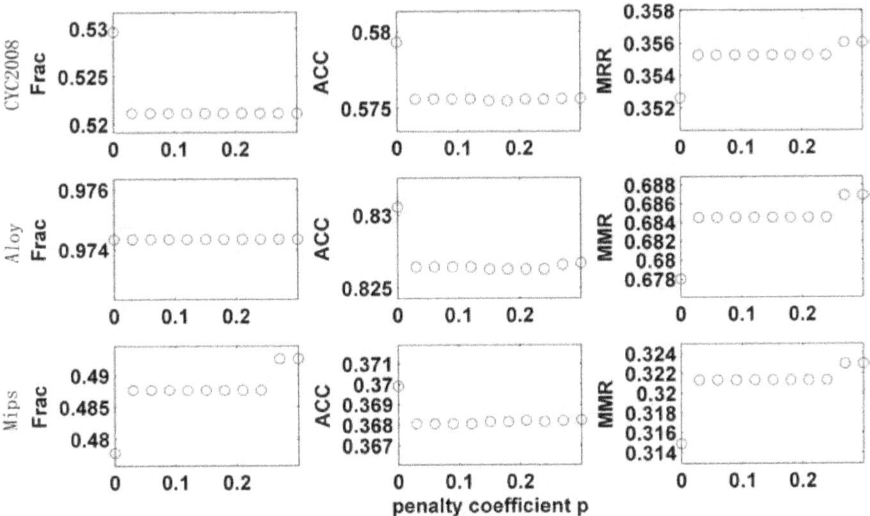

Fig. 6. The three core measurement plots as a function of penalty term p under the Gavin PPI network. The three rows of the figures (on the left, middle and right) are corresponding to the CYC2008, Aloy and Mips reference protein complexes datasets.

interval [0, 0.3], the new proposed method always achieves good results under the tested PPI networks. Hence, we set parameter p to 0.1 in performing simulation under all of the considered PPI datasets.

3.5. *Effect of PCC threshold on prediction performance*

In the IPC-PRIC algorithm, the prediction accuracy of the method is closely related to the PCC threshold. To investigate the effect of PCC threshold on performance of our method, we evaluate the prediction measurements by setting the PCC threshold from 0.90 to 1 with a step 0.01. When the PCC threshold is set to 1, this corresponds to the original network. Figures 7–9 show the three core measurement plots as a function of the PCC threshold when the PCC threshold varies from 0.9 to 0.99 with a step 0.01 for the Gavin, Collins and Krogan_extended networks, respectively. According to the results shown in the three figures, with an increase in the PCC threshold, the prediction accuracy is increased for both networks because with an increase in the PCC threshold, the added links in the constructed network will be more reliable, which contributes to the increase in ACC measurement. As for the MMR measurement, it also increases slightly with an increase in the PCC threshold. By contrast, the matched number of predicted protein complexes is not increased

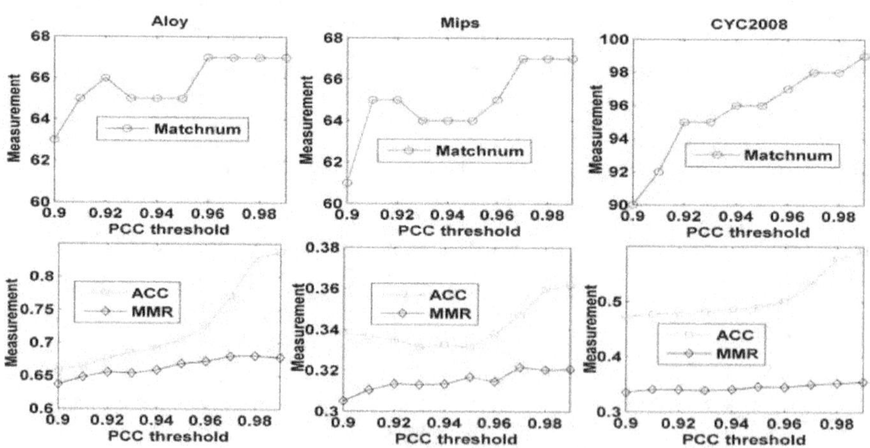

Fig. 7. The three core measurements as a function of PCC threshold under the Gavin PPI network. The three columns of subfigures are corresponding to the Aloy, Mips and CYC2008 reference protein complexes datasets.

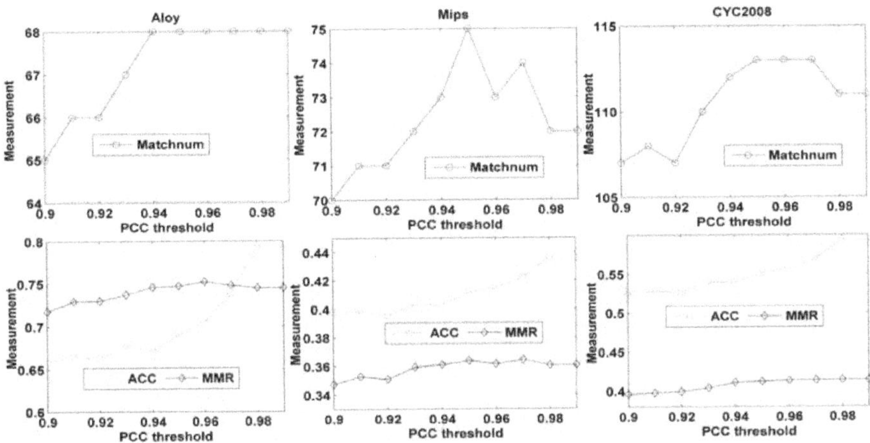

Fig. 8. The three core measurements as a function of PCC threshold under the Collins PPI network. The three columns of subfigures are corresponding to the Aloy, Mips and CYC2008 reference protein complexes datasets.

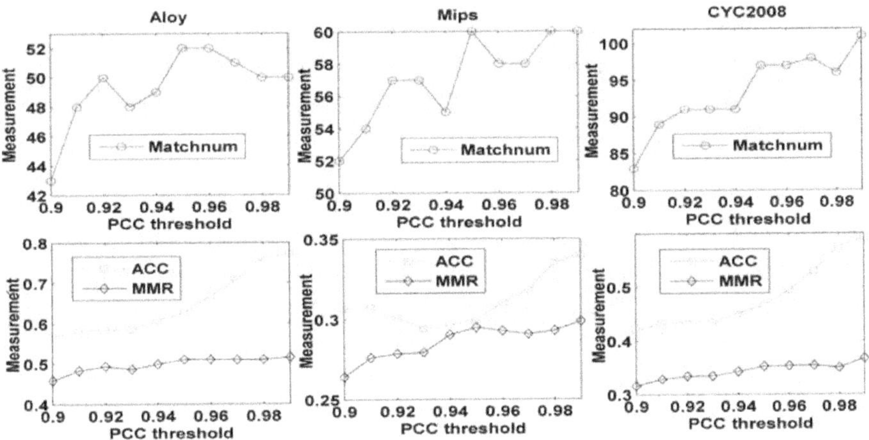

Fig. 9. The three core measurements as a function of PCC threshold under the Krogan_extended PPI network. The three columns of subfigures are corresponding to the Aloy, Mips and CYC2008 reference protein complexes datasets.

monotonically when the PCC threshold varies from 0.9 to 0.99; it depends on the PPI network and reference protein complexes we choose. Overall, the matched number is higher when the PCC threshold is set to be larger than 0.94.

4. Conclusions

Predicting protein complexes from PPI networks is a hot topic in the post genome era and many computational methods have been developed to predict protein complexes for the PPI network. However, most of these methods are based on the original network only and the incomplete data hinder the prediction of protein complexes.

In this research, we constructed a refined PPI network based on time series gene expression data, and then introduced a new way to improve the prediction accuracy of the protein complexes by incorporating topology properties and GO annotation information. The compactness of the subnetwork is characterized by the number of three node cliques and GO functional semantic similarity under BP sub-annotation information. Based on the refined network and new evaluation of the compactness of the subnetwork, we developed a new method called IPC-RPIN for protein complex prediction.

To evaluate the performance of the proposed strategy, we validated the new method on three test PPI networks under three reference protein complex datasets and compared it with four state-of-the-art methods. The simulation results demonstrated that the performance of the new proposed strategy is competitive in predicting protein complexes and that adding the GO annotation information will help to increase the prediction accuracy of protein complexes.

Although the new strategy performs well in the detection of protein complexes under the PPI network, the new network obtained by link prediction is still being refined. False-positive and negative links in the networks have not been considered, and the computational complexity is relatively large. Therefore, in the future, we will try to design a new parallel method in predicting unrevealed links and filtering the noise underlying the PPI network.

Acknowledgments

This work was partly supported by the National Natural Science Foundation (Nos. 61802125, 11626102 and 61672388), the Natural Science Foundation of Jiangxi Province (Nos. 20181BAB202006 and 20161BAB211022), and the Science Foundation of Wuhan Institute of Technology (Project No. K201746).

References

1. Spirin V, Mirny LA, Protein complexes and functional modules in molecular networks, *Proc Natl Acad Sci USA* **100**(21):12123–12128, 2003.
2. Rigaut G, Shevchenko A, Rutz B, Wilm M, Mann M, Séraphin B, A generic protein purification method for protein complex characterization and proteome exploration, *Nat Biotechnol* **17**(10):1030–1032, 1999.
3. Fields S, Song O, A novel genetic system to detect protein-protein interactions, *Nature* **340**(6230):245–246, 1989.
4. Uetz P, Giot L, Cagney G, Mansfield TA, Judson RS, Knight JR *et al.*, A comprehensive analysis of protein-protein interactions in saccharomyces cerevisiae, *Nature* **403**(6770):623–627, 2000.
5. Gavin AC, Bösche M, Krause R, Grandi P, Marzioch M, Bauer A *et al.*, Functional organization of the yeast proteome by systematic analysis of protein complexes, *Nature* **415**(6868):141–147, 2002.
6. Ito T, Chiba T, Ozawa R, Yoshida M, Hattori M and Sakaki Y, A comprehensive two-hybrid analysis to explore the yeast protein interactome, *Proc Natl Acad Sci USA* **98**(8):4569–4574, 2001.
7. Dongen SV, Graph Clustering by flow simulation, PhD Thesis, University of Utrecht, 2000.
8. Bader GD, Hogue CW, An automated method for finding molecular complexes in large protein interaction networks, *BMC Bioinform* **4**(1):2, 2003.
9. Ulitsky I, Shamir R, Identification of functional modules using network topology and high-throughput data. *BMC Systems Biol* **1**(1):8, 2007.
10. Chua HN, Ning K, Sung WK, Leong HW, Wong L, Using indirect protein-protein interactions for protein complex prediction, *J Bioinform Comput Biol* **6**(3):435–466, 2008.
11. Wu M, Li XL, Kwoh CK, Ng SK, A core-attachment based method to detect protein complexes in PPI networks, *BMC Bioinform* **10**(1):169, 2009.
12. Liu GM, Wong L, Chua HN, Complex discovery from weighted PPI networks, *Bioinform* **25**(15):1891–1897, 2009.
13. Nepusz T, Yu HY, Paccanaro A, Detecting overlapping protein complexes in protein-protein interaction networks, *Nat Meth* **9**(5):471–472, 2012.
14. Lei X, Ding Y, Fujita H *et al.*, Identification of dynamic protein complexes based on fruit fly optimization algorithm, *Knowl-Based Syst* **105**:270–277, 2016.
15. Bandyopadhyay S, Ray S, Mukhopadhyay A, Maulik U, A multiobjective approach for identifying protein complexes and studying their association in multiple disorders, *Algorithms Mol Biol* **10**(1):24, 2015.
16. Cao B, Luo J, Liang C, Wang S, Song D, MOEPGA: A novel method to detect protein complexes in yeast proteinCprotein interaction networks based

on multiobjective evolutionary programming genetic algorithm, *Comput Biol Chem* **58**:173–181, 2015.

17. Wu M, Ou-Yang L, Li XL, Protein complex detection via effective integration of base clustering solutions and co-complex affinity scores, *IEEE/ACM Trans Comput Biol Bioinform* **14**(3):733–739, 2017.

18. Zaki N, Efimov D, Berengueres J, Protein complex detection using interaction reliability assessment and weighted clustering coefficient, *BMC Bioinform* **14**:163, 2013.

19. Peng W, Wang J, Zhao B, Wang L, Identification of protein complexes using weighted PageRank-Nibble algorithm and core-attachment structure, *IEEE/ACM Trans Comput Biol Bioinform* **12**(1):179–192, 2015.

20. Chen BL, Shi JH, Zhang SG, Wu FX, Identifying protein complexes in protein-protein interaction networks by using clique seeds and graph entropy, *Proteomics* **13**(2):269–277, 2012.

21. Shen X, Yi L, Jiang X, Zhao Y, Hu X, He T, Yang J, Neighbor affinity based algorithm for discovering temporal protein complex from dynamic PPI network, *Methods* **110**:90–96, 2016.

22. Kouhsar M, Zare-Mirakabad F, Jamali Y, WCOACH: Protein complex prediction in weighted PPI networks, *Genes Genetic Syst* **90**(5):317–324, 2015.

23. Keretsu S, Sarmah R, Weighted edge based clustering to identify protein complexes in proteinCprotein interaction networks incorporating gene expression profile, *Comput Biol Chem* **65**:69–79, 2016.

24. Zhang W, Zou X, A new method for detecting protein complexes based on the three node cliques, *IEEE/ACM Trans Comput Biol Bioinform* **12**(4):879–886, 2015.

25. Wang J, Li M, Deng Y, Pan Y, Recent advances in clustering methods for protein interaction networks, *BMC Genom* **11**(3):S10, 2010.

26. Srihari S, Leong, HW, A survey of computational methods for protein complex prediction from protein interaction networks, *J Bioinform Comput Biol* **11**(2):1230002, 2013.

27. Ji J, Zhang A, Liu C, Quan X, Liu Z, Survey: Functional module detection from protein-protein interaction networks, *IEEE Trans Knowl Data Eng* **26**(2):261–277, 2014.

28. Ou-Yang L, Wu M, Zhang XF, Dai DQ, Li XL, Yan H, A two-layer integration framework for protein complex detection, *BMC Bioinform* **17**(1):100, 2016.

29. Meng J, Zhang X, Luan Y, Global propagation method for predicting protein function by integrating multiple data sources, *Current Bioinform* **11**(2):186–194, 2016.

30. Zhang W, Xu J, Li Y, Zou X, Detecting essential proteins based on network topology, gene expression data and gene ontology information, *IEEE/ACM Trans Comput Biol Bioinform* **15**(1):109–116, 2018.

31. Yong CH, Liu G, Chua HN *et al.*, Supervised maximum-likelihood weighting of composite protein networks for complex prediction, *BMC Syst Biol BioMed Central* **6**(2):S13, 2012.

32. Wu M, Xie Z, Li X *et al.*, Identifying protein complexes from heterogeneous biological data, *Proteins: Struct Funct Bioinform* **81**(11):2023–2033, 2013.

33. Zhang Y, Lin H, Yang Z, Wang J, Li Y, Xu B, Protein complex prediction in large ontology attributed protein-protein interaction networks, *IEEE/ACM Trans Comput Biol Bioinform* **10**(3):729–741, 2013.

34. Mukhopadhyay A, Ray S, De M, Detecting protein complexes in a PPI network: A gene ontology based multi-objective evolutionary approach, *Mol Biosyst* **8**(11):3036–3048, 2012.

35. Hanna EM, Zaki N, Amin A, Detecting protein complexes in protein interaction networks modeled as gene expression biclusters, *PLoS One* **10** (12):e0144163, 2015.

36. Ou-Yang L, Dai DQ, Zhang XF, Protein complex detection via weighted ensemble clustering based on Bayesian nonnegative matrix factorization, *PLoS One* **8**(5):e62158, 2013.

37. Zhang Y, Lin H, Yang Z, Jian W, Liu Y, Sang S, A method for predicting protein complex in dynamic ppi networks, *BMC Bioinform* **17**(7):229, 2016.

38. Ou-Yang L, Dai DQ, Li XL, Wu M, Zhang XF, Yang P, Detecting temporal protein complexes from dynamic protein-protein interaction networks, *BMC Bioinform* **15**(1):335, 2014.

39. Li M, Meng X, Zheng R, Wu FX, Li Y, Pan Y, Wang J, Identification of protein complexes by using a spatial and temporal active protein interaction network, *IEEE/ACM Trans Comput Biol Bioinform*, doi: 10.1109/TCBB.2017.2749571, 2017.

40. Ma W, McAnulla C, Wang L, Protein complex prediction based on maximum matching with domain-domain interaction, *Biochimica et Biophysica Acta (BBA)-Proteins and Proteomics* **1824**(12):1418–1424, 2012.

41. Sprinzak E, Sattath S, Margalit H, How reliable are experimental protein-protein interaction data? *J Mol Biol* **327**(5):919–923, 2003.

42. Zhang W, Xu J, Li Y, Zou X, A new two-stage method for revealing missing parts of edges in protein-protein interaction networks, *PLoS One* **12**(5): e0177029, 2017.

43. Li M, Ni P, Chen X, Wang J, Wu F, Pan Y, Construction of refined protein interaction network for predicting essential proteins, *IEEE/ACM Trans Comput Biol Bioinform* 2017, doi: 10.1109/TCBB.2017.2665482.

44. Wang JZ, Du Z, Payattakool R, Yu PS, Chen CF, A new method to measure the semantic similarity of GO terms, *Bioinformatics* **23**(10):1274–1281, 2007.
45. Pu S, Wong J, Turner B, Cho E, Wodak SJ, Up-to-date catalogues of yeast protein complexes, *Nucl Acids Res* **37**(3):825–831, 2008.
46. Pagel P, Kovac S, Oesterheld M, Brauner B, Dunger-Kaltenbach I, Frishman G *et al.*, The MIPS mammalian protein-protein interaction database, *Bioinform* **21**(6):832–834, 2004.
47. Aloy P, Böttcher B, Ceulemans H, Leutwein C, Mellwig C, Fischer S *et al.*, Structure-based assembly of protein complexes in yeast, *Science* **303** (5666):2026–2029, 2004.
48. Tu BP, Kudlicki A, Rowicka M, McKnight SL, Logic of the yeast metabolic cycle: Temporal compartmentalization of cellular processes, *Science* **310** (5751):1152–1158, 2005.
49. Brohée S, Van Helden J, Evaluation of clustering algorithms for protein-protein interaction networks, *BMC Bioinform* **7**:488, 2006.

Wei Zhang received B.S. degree in China Three Gorges University in 2008, and received M.S. degree and the Ph.D. degree in Computational Mathematics from Wuhan University, Wuhan, China, in 2011 and 2014, respectively. He is currently a lecturer at the School of Science, East China Jiaotong University. His research interests include Computational Systems Biology, Complex Networks and Computational Intelligence.

Jia Xu received M.S. degree in Microelectronics from South China Normal University, Guangzhou, China, in 2012. She is currently a laboratory assistant at the School of Mechatronic Engineering, East China Jiaotong University. Her research interests include Microelectronics and Bioinformatics.

Yuanyuan Li received B.S. and M.S. degrees in Wuhan University in 2002 and 2005, respectively. She is currently working toward the Ph.D. degree at the School of Mathematics and Statistics, Wuhan University and she is also a lecturer at the School of Science at Wuhan Institute of Technology in Wuhan, China. Her research interests include Bioinformatics and System Biology.

Xiufen Zou received the B.S. and M.S. degrees in Computational Mathematics and the Ph.D. degree in Computer Science from Wuhan University, Wuhan, China, in 1986, 1991, and 2003, respectively. She is currently a full professor at the School of Mathematics and Statistics, Wuhan University. Her current research interests include Computational Intelligence and its Applications in Complex Biological Systems.

Part II
Host-Pathogen Interaction Prediction

CHAPTER 5

Progress in Computational Studies of Host–Pathogen Interactions[a]

Hufeng Zhou[*,‡], Jingjing Jin[†] and Limsoon Wong[†,§]

*NUS Graduate School for Integrative Sciences & Engineering
National University of Singapore, Singapore 117456*

*†School of Computing, National University of Singapore
Singapore 117417*
‡zhouhufeng@nus.edu.sg
§wongls@comp.nus.edu.sg

Host–pathogen interactions are important for understanding infection mechanism and developing better treatment and prevention of infectious diseases. Many computational studies on host–pathogen interactions have been published. Here, we review recent progress and results in this field and provide a systematic summary, comparison and discussion of computational studies on host–pathogen interactions, including prediction and analysis of host–pathogen protein–protein interactions; basic principles revealed from host–pathogen interactions; and database and software tools for host–pathogen interaction data collection, integration and analysis.

Keywords: Host–pathogen interaction; protein–protein interaction; PPI; infectious diseases.

1. Introduction

Infectious diseases are among the leading causes of death worldwide. Host–pathogen interactions are crucial for better understanding of the

---

[a]This article was previously published in *Journal of Bioinformatics and Computational Biology*. Vol: 11, No. 2 (2013), 1230001 (26 pages).

mechanisms that underline infectious diseases and for developing more effective treatment and prevention measures. While host–pathogen interactions take many forms, in this review, we concentrate on protein–protein interactions (PPIs) between a pathogen and its host. This review consists of the following parts: (i) host–pathogen PPIs prediction; (ii) basic principles derived from the analysis of known host–pathogen PPIs; (iii) host–pathogen PPIs analysis and assessment; and (iv) host–pathogen interaction data collection and integration.

Several approaches have been proposed to computationally predict host–pathogen PPIs. There has also been progress on analyzing and assessing the quality of the inferred host–pathogen PPIs. This has led to cataloging of PPI data that can be further analyzed to understand the impact of these interactions (especially on the host) and to decipher underlying disease mechanisms. Approaches developed for predicting host–pathogen PPIs can be broadly categorized into homology-based,[1–5] structure-based,[6–8] domain–motif interaction-based approaches,[9,10] as well as machine-learning–based approaches.[11–13] These approaches can also be combined and used together in some studies to improve prediction performance. These approaches are reviewed in Sec. 2.

An analysis of experimentally verified as well as manually curated host–pathogen PPIs have led to a number of observations. These observations include the topological properties of targeted host proteins and structural properties of host–pathogen PPI interfaces. These observations are discussed in Sec. 3.

Approaches for assessing and analyzing host–pathogen PPIs can be categorized into assessment based on gold standard PPIs,[6,7,10–13] functional information analysis [Gene Ontology (GO),[5–8,10,11] pathways,[5,10,14,15] gene expression data,[2,5,6] RNA interference data[7,8,10–13]], localization information analysis (protein subcellular localization,[1–5] co-localization of host and pathogen proteins[7,8]), related experimental data analyses[8,11,13] and biological case studies and explanations.[2–4,6–8,12] Some of these assessment approaches can also be used as filtering strategies for pruning host–pathogen PPI prediction results. These approaches and the outcome of the analysis are reviewed in Sec. 4.

Host–pathogen PPIs curated from primary literature are usually facilitated by text-mining techniques.[16,17] With more host–pathogen PPI data available from literature curation and experiments, there are strong needs for data collection and integration facilities that can provide comprehensive storage, convenient access and effective analyses of the integrated host–pathogen interaction data. The development of software and database tools dedicated to host–pathogen interaction data collection, integration and analysis are also very prominent. Integration of host–pathogen interaction data is not confined to PPI data. Other related data — like pathogen virulence factors, human-diseases–related genes, sequence and homology information, pathway information, functional annotations, diseases information, literature sources, etc. — are also being integrated into several databases. These databases[16–25] and softwares[26] are reviewed in Sec. 5.

2. Host–Pathogen PPIs Prediction

Host–pathogen PPIs play an important role between the host and pathogen, which may be crucial in the outcome of an infection and the establishment of disease. Unfortunately, experimentally verified interactions between host and pathogen proteins are currently rather limited for most host–pathogen systems. This has motivated a number of pioneering works on computational prediction of host–pathogen PPIs. These works can be roughly categorized into modeling approaches based on sequence homology, protein structure, domain and motif and approaches based on machine learning. These pioneering works are reviewed and discussed below.

2.1. *Homology-Based Approach*

The homology-based approach is a conventional way for predicting intra-species PPIs. Many studies have also adopted this strategy for predicting host–pathogen PPIs, which are inter-species PPIs. The basic hypothesis of the homology-based approach is that the interaction between a pair of proteins in one species is expected to be conserved in related species.[27] This is a reasonable hypothesis as a pair of homologous proteins descend from the same ancestral pair of interacting proteins and is expected to inherit the structure and function and, thus, interactions of the ancestral proteins. Therefore, the basic procedure of the homology-based approach for intra-species PPI prediction is (i) starting from a known PPI (the template PPI) in some source species, (ii) determining in the target species the homologs (x', y') of the two proteins (x, y) in the template PPIs, and (iii) predicting that the two homologs (x', y') interact in the target species. This approach is generally adapted to the inter-species scenario of host–pathogen PPI prediction by (i) starting from a known PPI (the template PPI) in some source species, (ii) determining in the host a homolog (x') and in the pathogen a homolog (y'), respectively, of the two proteins (x, y) in the template PPI and (iii) predicting that (x', y') interact.

The main advantages of the homology-based approach to host–pathogen PPI prediction are its simplicity and its apparent biological basis. Since the data required for performing the prediction are only the template PPIs and protein sequences, this approach is scalable and can be applied to many different host–pathogen systems. The homology-based approach can be used alone[1–4] or in combination with other methods[5] in predicting host–pathogen PPIs. The investigated host–pathogen systems in past studies include *Homo sapiens–Plasmodium falciparum*,[1,2,5] *H. sapiens–Heliobacter pylori*,[3] *phage T4–Escherichia coli*,[4] *phage lambda–E. coli*,[4] *H. sapiens–E. coli*,[4] *H. sapiens–Salmonella enterica*,[4] *H. sapiens–Yersinia pestis*,[4] etc. The template PPIs used in the prediction can also be very different. The commonly used template PPIs are from DIP,[28] iPfam,[29] MINT,[30] HPRD,[31] Reactome,[32] IntAct,[33] etc.

There is an inherent weakness in the homology-based approach. Basically, in a real biological process, such as infection, the two proteins in a predicted PPI may actually have little opportunity to be present together. Consequently, host–pathogen PPIs predicted solely on the homology basis, without considering other biological properties of the proteins involved, may not be very reliable. Additional information should be used to increase the accuracy of the prediction. For example, extracellular localization and transmembrane regions are used in pruning[4] or constraining the predictions.[3] Also, a pathogen (e.g. *P. falciparum*) may infect different organs at different stages of the pathogen's life cycle. Thus, filtering by tissue-specific gene expression data may also improve prediction reliability.[2] Indeed, recognizing this weakness in the homology-based approach, Wuchty[5] has proposed filtering PPIs predicted by the homology-based approach using a random-forest classifier trained on sequence compositional characteristics of known PPIs, as well as by gene expression and molecular characteristics. This results in a significantly smaller set of putative host–pathogen PPIs, which are claimed to be of higher quality than the original set of predicted PPIs.

2.2. *Structure-Based Approach*

When a pair of proteins have structures that are similar to a known interacting pair of proteins, it is reasonable to believe that the former are likely interacting in a way that is structurally similar to the latter. In accordance to this hypothesis, several works have used structural information to identify the similarity between query proteins (i.e. proteins in the pathogen and host) and template PPIs (i.e. known interacting protein pairs) and infer that those host–pathogen protein pairs that match some template PPIs are interacting.

2.2.1. *Comparative Modeling*

Prediction by comparative modeling is a representative structure-based approach. For example, in Davis *et al.*,[6] an automated pipeline for large-scale comparative protein structure modeling, MODPIPE, is applied to model the structure of host and pathogen proteins based on their sequences and corresponding template structures. Given the computed model of a protein, the SCOP[34] superfamilies that the protein belongs to are identified. A database of protein structural interfaces, PIBASE, is then scanned. If a SCOP superfamily of a host protein and a SCOP superfamily of a pathogen protein are both involved in the same PIBASE[35] protein structural interface, then the host protein and the pathogen protein are predicted as a putative PPI.

Query proteins that lack structural templates cannot be modeled in the above process. In this case, template interactions in alternative databases (e.g. IntAct) are considered by Davis *et al.*[6] Specifically, a pair of host and pathogen proteins are predicted to interact if at least 50% of each of the two protein sequences are similar to some member proteins of a template complex in IntAct and the joint sequence identity ($\sqrt{Sequence\ Identity1 * Sequence\ Identity2}$) is at least 80%. These predictions, which are conducted without structural information, form a very small portion of the total number of putative PPIs, because of the stringent joint threshold. Each prediction is further followed by a series of assessments and filtering (biological and network filters), which results in a significant reduction of potential host–pathogen PPIs by several order of magnitudes.

2.2.2. *Structural Similarity*

Structural similarity can also be analyzed using the Dali database.[36] This strategy has been adopted to predict *H. sapiens*–HIV PPIs,[8]

H. sapiens–DENV PPIs[7] and *A. aegypti*–DENV PPIs.[7] Dali calculates structural similarity score by comparing the 3D structural coordinates of two PDB entries.[7] To predict the *H. sapiens*–HIV and *H. sapiens*–DENV PPIs, structurally similar pathogen (HIV, DENV) and host (*H. sapiens*) proteins are first determined using Dali. Then, under the assumption that pathogen proteins having similar structure to host proteins are likely to participate in the similar set of PPIs (*H. sapiens* PPI dataset from HPRD[31]) that those matched host proteins participate in, the pathogen proteins are directly mapped to their high-similarity matches within the host intra-species PPI network to predict the host–pathogen PPIs.[7,8] The same structural similarity prediction method has been applied to identify orthologs between *Drosophila melanogaster* and *Aedes aegypti* and map *D. melanogaster*–DENV PPIs to predict *A. aegypti*–DENV PPIs[7] — the host–pathogen PPIs between DENV and its real insect host. The accuracy of this prediction method depends on the performance of Dali in determining structurally similar pathogen and host proteins. The availability of pathogen and host protein structures and the quality of host intra-species PPI data also have a significant influence on prediction results.

2.3. *Domain and Motif Interaction-Based Approach*

Domains are basic building blocks determining the structure and function of proteins and they play specialized role in mediating the interaction of proteins with other molecules.[37] Some studies have proposed predicting host–pathogen PPI based on domain–domain interaction (DDI)[9] and motif–domain interaction.[10]

2.3.1. *Domain–Domain Interaction-Based Approach*

Dyer *et al.*[9] predict host–pathogen PPIs in the *H. sapiens*–*P. falciparum* system by integrating known intra-species PPIs with domain profiles based on an association method (sequence-signature algorithm) proposed by Sprinzak and Margalit.[38] Specifically, domains are first identified by InterProScan[39] in each interacting protein in the intra-species PPIs. Then, the probability $P(d, e)$ that two proteins containing a specific pair of domains (d, e) would interact is estimated for each pair of domains in the Bayesian manner. Finally, given a pair of host–pathogen proteins, their probability of interaction is estimated by a naive combination

$(= 1 - \prod_i \prod_j (1 - P(d_i, e_j)))$ of the probabilities from each pair of domains (d_i, e_j) contained in the pair of proteins.[9]

At around the same time, Kim *et al.*[40] predict *H. sapiens–H. pylori* PPIs using the PreDIN[41] and PreSPI[42] algorithms, which are also based on domain information. The domain annotation used in this work is done by InterProScan as well. However, in contrast to Dyer *et al.*,[9] which is based on estimating the probability of an individual pair of domains being associated with protein interactions and naively combining these probabilities, PreDIN and PreSPI directly estimate the probability of domain combination pairs being associated protein interactions.

2.3.2. *Motif–Domain Interaction-Based Approach*

Some protein interactions are mediated not by interactions between domains but by interactions between a domain in one protein and a short linear motif (SLiM) in the other protein.[43,44] As viral pathogens typically have a compact genome, they have few domains. It is reasonable to postulate that their interaction with host proteins are likely to be mediated by Domain–SLiM interactions. For example, since HIV-1 proteins have few domains, Evans *et al.*[10] predicted *H. sapiens–*HIV-1 PPIs based on the interactions between short eukaryotic linear motifs (ELMs) and human protein counter domains (CDs).

Evans *et al.* use the ELM resource[45] to determine ELMs contained in human and HIV-1 proteins and PROSITE[46] to determine domains in human proteins. Then starting from a template human PPI (x, y) where protein x contains a ELM (E) and protein y a counter domain (CD), proteins in HIV-1 that contain the ELM (E) are predicted to form host–pathogen PPIs with the human protein y. Notably, Evans *et al.* point out that the human protein x is expected to compete with these HIV-1 proteins for interacting with y, and that this competition should be considered as another form of host–pathogen interaction.

2.4. *Machine-Learning–Based Approach*

Both supervised[11,12] and semi-supervised[13] learning frameworks have also been used in predicting host–pathogen PPIs. A considerable amount of interacting and non-interacting pairs are usually needed by these machine learning algorithms to produce good classifiers. For example, Tastan *et al.*[11] and Qi *et al.*[13] obtain curated *H. sapiens–*HIV PPIs from the

"HIV-1, human protein interaction database",[19] while Dyer *et al.*[12] compile *H. sapiens*–HIV PPIs from other sources including BIND,[47] DIP,[28] IntAct[33] and Reactome.[32] Supervised learning framework has first been attempted using a Random Forest (RF)[11] classifier with 35 selected features, including GO similarity, graph properties of the human interactome, ELM-ligand, gene expression, tissue feature, sequence similarity, post-translational modification similarity to neighbor, HIV-1 protein type, etc. In another work,[12] a Support Vector Machine (SVM) is used with linear kernel and features such as domain profiles, protein sequence *k*-mers and properties of human proteins in the human interactome.

The performance of supervised learning algorithms is limited by the availability of truly interacting proteins. However, there are a lot of protein pairs that have a known association between themselves which may not be a confirmed direct interaction.[13] In order to exploit the availability of these data, Qi *et al.*[13] try a semi-supervised learning approach.

The semi-supervised approach of Qi *et al.*[13] uses the same training data (collected by Fu *et al.*[19]) as the supervised approach of Tastan *et al.*,[11] who use only physical PPIs with keywords "interact," "bind," etc. for training. However, Qi *et al.*[13] use only a subset of the physical PPIs used by Tastan *et al.*[11] This subset consists of 158 expert-annotated *H. sapiens*–HIV PPIs and is labeled as positive training data. The remaining PPIs from Fu *et al.*[19] are used as "partial positive" training data. This is because Qi *et al.* find that many of the PPIs — even those with keywords "interact," "bind," etc. — are not well agreed by experts.[13] Moreover, only 18 of the 35 attributes used by Tastan *et al.* are used by Qi *et al.* Despite using fewer attributes, the separation of the PPI training data into definite known positive interactions and partial positives helps Qi *et al.* achieve a higher performance than Tastan *et al.*

An important weakness of these approaches based on machine learning is that the features used by them — e.g. the domain profile feature[12] and the HIV-1 protein type feature[11] — are not easy to understand, especially with respect to their biological basis. Another weakness is the limitation of training data. For example, the use of machine learning approaches in the context of host–pathogen PPI prediction has so far been applied in the *H. sapiens*–HIV system because known host–pathogen PPIs are not available in other host–pathogen systems on a sufficiently large scale.

3. Basic Principles of Host–Pathogen Interaction

Some basic principles derived from the analysis of experimentally verified or manually curated host–pathogen PPIs are discussed in this section. These principles either have been reported and confirmed by several works or have high potential to be applied in future works on host–pathogen interactions.

3.1. *Topological Properties of Targeted Host Proteins*

Calderwood *et al.*[48] have generated 44 intra-species Epstein-Barr virus (EBV) PPIs and 173 inter-species *H. sapiens*–EBV PPIs using a stringent and systematic two-hybrid system. They observe that the degree (in the human interactome) of human proteins involved in *H. sapiens*–EBV PPIs are significantly higher than randomly selected human proteins. Thus, these targeted human proteins are enriched with hubs (i.e. proteins with high degree in the human interactome).

Moreover, Calderwood *et al.*[48] also report that the minimum number of steps (in terms of PPI edges) between a targeted human protein and a reachable protein in the network is, on average, smaller than that of randomly picked human proteins. Thus the EBV-targeted human proteins have relatively shorter paths to other proteins in the human interactome.[48]

Dyer *et al.*[49] have also analyzed the topological properties of pathogen-targeted host proteins using much larger datasets. The inter-species host–pathogen PPI and intra-species human PPI datasets studied are integrated from primary literature[48] and 7 databases.[28,30–33,47,50] This integrated host–pathogen PPI dataset contains 10,477 experimentally detected and manually curated host–pathogen PPIs, covering 190 pathogens (most of which are viruses), while the integrated human PPI dataset contains 75,457 experimentally verified PPIs.[49] The result reveals that proteins interacting with viral and bacterial pathogen groups tend to have higher degrees (hubs), which confirms one of the observations of Calderwood *et al.*,[48] and higher betweenness centrality (bottlenecks).

Dyer *et al.* also analyze the physical interaction network between human and three bacterial pathogens (*Bacillus anthracis*, *Francisella tularensis* and *Y. pestis*) generated from a modified two-hybrid assay (liquid-format mating).[51] The analyses show again that pathogens preferentially interact with hubs and bottlenecks in the human interactome.[51] Zhao *et al.*[15] have similarly confirmed that hubs are more likely to be targeted by viruses in studying human–virus PPIs and human signal transduction pathways.

3.2. *Structural Properties of Host–Pathogen PPIs*

Franzosa and Xia[52] report a significant overlap between exogenous (i.e. host–pathogen) and endogenous (i.e. within-host) interfaces of PPIs, suggesting interface mimicry as a possible pathogen strategy to evade immune system detection and to hijack host cellular machinery. The exogenous interactions represent clear cases of horizontal gene transfer between the virus and host.[52] The acquisition of viral protein sequences from hosts are also observed and discussed by Rappoport and Linial.[53]

Comparing with endogenous interfaces, exogenous interfaces tend to be smaller, indicating that the viral genome is under intense selection to reduce its size compared to the host genome.[52] There is a similar observation in another work[53] that viral proteins are noticeably shorter than their corresponding host counterparts, which may result from acquiring only host gene fragment, eliminating internal domain and shortening domain linkers.

Interestingly, Franzosa *et al.*[52] find that virus-targeted interfaces tend to be "date"-like. That is, they are transiently used by different endogenous binding partners at different times and, on average, they utilize more human binding partners than generic endogenous interfaces. This finding is supported by functional enrichment among the mimicked endogenous binding partners for the GO term "Regulation of Biological Process",[52] since proteins involved in biological regulation usually have transient binding with other proteins. This may also partially explain the topological property that targeted host proteins tend to be hubs in the host interactome,[48] because the proteins having date-like interfaces tend to interact with many proteins and appear as hubs in intra-species PPI networks.

Lastly, an analysis of residues involved in exogenous and endogenous interfaces shows that exogenous interfaces are likely to be less conserved then endogenous interfaces.[48]

4. Analysis and Assessment of Host–Pathogen PPIs

Analysis of host–pathogen PPI datasets is essential both for developing better prediction approaches and applying the host–pathogen PPI datasets in the subsequent studies. Assessment and analysis of host–pathogen PPI datasets can be conducted directly using (i) gold standard host–pathogen PPIs or indirectly using (ii) functional information, (iii) localization information, (iv) related experimental data, (v) biological explanation of selected examples, etc.

4.1. *Assessment Based on Gold Standard*

Known truly interacting host–pathogen PPI data (gold standard) are available for a few pathogens. The "HIV-1, Human Protein Interaction database"[19] contains a considerable number of *H. sapiens*–HIV PPIs. A substantial number of host–pathogen PPIs (mainly *H. sapiens*–HIV PPIs) can also be found in other databases including BIND,[47] DIP,[28] IntAct[33] and Reactome.[32] Therefore, in the case of *H. sapiens*–HIV PPIs, a fairly large gold standard dataset is available. For example, the "HIV-1, Human Protein Interaction database"[19] has been used in assessing predictions based on motif–domain interaction.[10] On the other hand, Davis *et al.*[6] have only managed to collect 33 host–pathogen PPIs from the literature to validate their predictions for 10 pathogen species. As another example, Doolittle and Gomez[7] have only managed to collect 3 PPIs from a public database[49] and 20 PPIs from the literature, and only 19 among these collected PPIs are specific to the *H. sapiens*–DENV-2 system that Doolittle and Gomez[7] have made predictions for. Although 9 of these 19 gold standard PPIs are present in the prediction results of Doolittle and Gomez,[7] the assessment has been badly hampered by the small size of the gold standard dataset.

4.2. *Analyses and Assessments Based on Functional Information*

4.2.1. *Gene Ontology*

Gene Ontology (GO) terms that are significantly enriched in the host proteins predicted to be targeted by pathogens can be used to evaluate the functional relevance of the predicted host–pathogen PPIs.[6] GO terms specific for human proteins involved in the immune system and for pathogen proteins involved in host–pathogen interactions can also be used to filter putative host–pathogen PPIs.[6]

Several tools can analyze GO term enrichment, including GOstat[54] used by Wuchty,[5] GO::TermFinder[55] used by Davis *et al.*,[6] Ontologizer[56] used by Tanstan *et al.*[11] and DAVID[57] used in many other studies.[7,8,10]

Specifically, Wuchty[5] analyzes the GO term enrichment of host proteins in predicted *H. sapiens–P. falciparum* PPIs and derives the 100 most enriched GO terms (in the Biological Process category) of host proteins. He finds that the pathogen may influence important signaling and regulation processes of the host through host–pathogen PPIs.[5] Tastan *et al.*[11] analyze the GO term enrichment of host proteins in predicted host–pathogen PPIs; they find that 31 GO terms in the Molecular Function category (e.g. transcription regulator, ligand-dependent nuclear receptor, MHC class I receptor and protein kinase C activities), 19 GO terms in the Biological Process category (e.g. immune system process and response to stimulus) and 14 GO terms in the Cellular Component category (e.g. membrane-enclosed lumen and plasma membrane) are significantly enriched. Enriched GO terms are identified similarly in several studies[7,8] and, results show consistency with viral infection. Similarly, enriched GO terms have also been analyzed for pathogen groups[49] and Conserved Protein Interaction Modules (CPIM)[51] among *H. sapiens–B. anthracis, H. sapiens–F. tularensis* and *H. sapiens–Y. pestis* protein interaction networks.

4.2.2. *Pathway Data*

An analysis of host–pathogen PPIs in the context of biological pathways provides a functional overview of the targeted host proteins, illuminates the mechanisms of a pathogen's obstruction on host pathways and serves as an important assessment of predicted host–pathogen PPIs. We first discuss some results derived from an analysis of the known host–pathogen PPIs using pathway data. Then we introduce some assessment strategies of predicted host–pathogen PPIs using pathways.

Balakrishnan *et al.*[58] analyze the PPI dataset from the "HIV-1, Human Protein Interaction database"[19] in the context of human signal transduction in the Pathway Interaction Database (PID)[59] and Reactome.[32] They discover that a majority of human pathways can potentially be targeted by *H. sapiens–HIV-1* PPIs. However, many alternative paths (starting and ending at the same proteins yet circumventing HIV-1 disrupted intermediate steps) to the HIV-1 targeted paths exist due to human network redundancy; and degradation and downregulation pathways are among the most highly targeted pathways. Singh *et al.*[14] and Zhao *et al.*[15] have also obtained similar results from analyzing the same pathway data: human signal transduction pathways derived from Pathway Interaction Database (PID)[59] and Reactome[32] and virus–host PPI data from

VirusMINT.[16] They find that 355 out of 671 pathways are targeted by at least one viral protein. Moreover, the majority of the pathways (268 out of 355) are targeted by more than one viral protein. In these 355 pathways, 413 proteins are targeted by 28 different viruses. Also, 95 of these 413 targeted host proteins are known drug targets.[14,15] However, proteins targeted by different viruses in each pathways are not necessarily the same. Zhao *et al.*[15] further report that centrally located proteins in merged networks of statistically significant pathways are hub proteins and are more frequently targeted by viruses.

Wuchty[5] analyzes both predicted and external (experimentally determined and structurally inferred) *H. sapiens–P. falciparum* PPIs using 184 manually curated pathways from PID.[59] He reports that both separate and combined sets of predicted and external PPIs target proteins which have a higher degree and which appear in more pathways.[5] For each pathogen protein, Wuchty[5] identifies pathways enriched with host proteins that are targeted by this pathogen protein using Fisher's exact test. He then constructs a bipartite matrix between pathogen proteins and their corresponding enriched host signaling pathways. Observation of the matrix reveals that the pathogen has many interactions with proteins in the TNF- and NF-kappa B pathways, which indicate the pathogen's obstruction of inflammatory response.[5] To evaluate host–pathogen PPIs predicted by the domain–motif interaction-based approach, KEGG pathway enrichment for HIV-1 proteins (ENV, NEF and TAT) targeted host proteins in the (experimentally verified and computationally predicted) inter-species host–pathogen PPIs are analyzed.[10] The enriched pathways include (i) immune system pathways such as T-cell and B-cell receptor signaling pathways, apoptosis, focal adhesion and toll-like receptor signaling pathways; (ii) disease pathways such as the colorectal cancer, leukemia and lung cancer pathways; and (iii) signal transduction processes.[10]

4.2.3. *Gene Expression Data*

Gene expression data are another important functional information source which have been widely used in the filtering, assessment and verification of host–pathogen PPIs. Tissue-specific and infection-related gene expression data are frequently used in host–pathogen studies. A pathogen like *P. falciparum* infects different human organs at different stages of its life cycle. So the expression data of different stages of its life cycle and *H. sapiens* tissue-specific gene expression data can be used simultaneously for pruning putative *H. sapiens–P. falciparum* PPIs.[2,5] For example, *P. falciparum* invades *H. sapiens* liver tissue during the sporozoite stage. The predicted host–pathogen PPIs are thus more likely to be real, if the corresponding human proteins are known to express in liver tissue and the corresponding pathogen proteins are known to express in the sporozoite stage. This filtering strategy has been adopted by several studies.[2,5] For the *H. sapiens–M. tuberculosis* system, human proteins expressed in lung tissue or bronchial epithelial cells and pathogen proteins upregulated in granuloma, pericavity or distal infection sites can be used for filtering purposes.[6] Moreover, pathogen genes involved in *M. tuberculosis* infections[60,61] and human genes involved in *M. tuberculosis*, *L. major* and *T. gondii* infections[62] can be compared with the pathogen and host proteins in predicted *H. sapiens–M. tuberculosis* PPIs as a useful assessment.[6]

4.2.4. *RNA Interference Data*

RNA interference (RNAi) is a natural process to specifically and selectively inhibit a targeted gene expression. Small interfering RNA (siRNA), short hairpin RNA (shRNA) and bi-functional shRNA are often used to mediate the RNAi effect. Some human proteins, when being silenced by genome-wide RNAi experiments, are found to be nonlethal to human cells but are essential for HIV replication. These human proteins may have high likelihood of interacting with HIV. Therefore, comparing the set of host proteins in predicted host–pathogen PPIs and the set of host proteins identified by RNAi experiments can be used as an assessment. We briefly list some examples below.

Several studies show that knocking down some host proteins by siRNA[63–65] or shRNAs[66] can impair HIV-1 infection or replication. Thus, those host proteins are essential for HIV-1 infection or replication. Therefore, they have higher possibility to interact HIV-1 proteins. This has been used as a filtering criterion[8] and assessment data[10–13] in several studies.

Three works[11–13] based on the machine learning approach for predicting *H. sapiens–*HIV PPIs use an siRNA dataset[64] to assess their prediction results. The assessment is conducted by examining the overlap between the human proteins targeted by the predicted PPIs and the proteins in the siRNA dataset.[64] Besides, Qi *et al.*[13] also combined four RNAi datasets[63–66] and conducted additional assessment in a similar way.

A five-way comparison has been conducted by Evans *et al.*[10] on five HIV-1 targeted human protein datasets — viz., (i) the human protein dataset targeted by PPIs predicted using the motif–domain interaction-based approach[10]; (ii) human protein dataset targeted by gold standard PPIs from the "HIV-1, Human Protein Interaction database"[19]; and (iii) human protein datasets from three genome-wide RNAi experiments.[63–65] Results show that genome-wide RNAi experiments match each other better than the interaction studies.[10] The matches between protein dataset (i) and the other four protein sets are significant but discrepancies are still observed.[10]

For the *H. sapiens*–DENV system, host protein datasets from two siRNA experiments in DENV infection[67,68] are available. They have also been used to refine *H. sapiens*–DENV PPI prediction result.[7]

4.3. *Pruning Based on Localization Information*

Localization information of pathogen and host proteins may relate to the possibility of their interactions. For extracellular pathogens, their extracellular or secretion proteins may have higher chance of interacting with host surface proteins rather than host nuclear proteins. For intracellular pathogens like viruses, co-localization of host and pathogen proteins may be one of the prerequisites for protein interactions. Several studies use this information to filter prediction results.

4.3.1. *Subcellular Localization of Host and Pathogen Proteins*

Since pathogen extracellular and secretion proteins, and proteins with translocational signals are more likely to interact with host extracellular or membrane proteins, such subcellular localization information are often used in pruning of predicted host–pathogen PPIs.[1–5] In connection with this, several tools are used in homology-based approaches[2–4] to predict protein subcellular localization.

4.3.2. *Co-Localization of Host and Pathogen Proteins*

As obligate intracellular pathogens, viruses do not have cellular structure or their own metabolism and are solely dependent on the host cell. Therefore, a viral protein and its host protein interaction targets are more likely to be co-localized. Several studies use this basic assumption to assess or filter predicted *H. sapiens*–HIV PPIs[8] and *H. sapiens*-DENV PPIs.[7] Similar information is also used as one of the selected features for classifiers in approaches based on machine learning for predicting *H. sapiens*–HIV PPIs.[11,13] The co-localization information of two proteins can be revealed through their shared GO terms in the Cellular Compartment category.

4.4. *Biological Explanation of Selected Examples*

An analysis of a specific PPI by explaining the underlining biological functions is not an effective assessment of predicted host–pathogen PPIs because such an analysis can cover only a small number of PPIs. However, it may facilitate a better understanding of that putative PPI and therefore promote subsequent experimental verification of that prediction. Explanation of the biological basis of some example PPIs from the whole dataset can be found in many studies.[2–4,6–8,12] Some of the specific examples may have literature or experimental supports, some lack direct literature support but have some indirect supports including structural information, homology to template PPIs, evidence from related experiments (gene expression and RNAi experiment data), etc. Explanation and identification of validated predictions also enhance the impact of the prediction methods; and this approach has been used in many studies.[4,6,12] For example, Dyer *et al.* discuss in detail the predicted *H. sapiens*–HIV PPIs[12] involving the HIV Dependency Factors[64] that have support in the literature.[12] To some extent, explanation of indirect evidence and clues enhances validity of the selected parts of the prediction results.[3,4,6–8] Predicted PPIs both with and without experimental verifications and PPIs involving hypothetical proteins are discussed and explained in Krishnadev and Srinivasan.[4] In another work, Tyagi *et al.*[3] also explain some examples of predicted *H. sapiens*–*H. pylori* PPIs through the structural point of view and discuss examples of PPIs involving membrane proteins, secreted proteins and hypothetical proteins.

4.5. *Assessment Through Related Experimental Data*

Some related experimental data turn out to be useful for assessing the targeted host proteins in host–pathogen PPIs. For example, during budding, host proteins may be incorporated into the virion.[69] Although some host proteins may be taken up by a budding virus accidentally, others are known to play crucial roles in viral life cycle and host–pathogen interaction. A dataset[69] on human protein presents in virion has also been used to filter predicted *H. sapiens*–HIV-1 PPIs.[8]

Qi *et al.*[13] and Tanstan *et al.*[11] use a human protein set hijacked by HIV-1 into its virion[70] to assess their predicted *H. sapiens*–HIV-1 PPIs. Specifically they examine the overlap between targeted human proteins in the predicted PPIs and the 314 human proteins in virion.[70] A large overlap suggests a satisfactory performance of the prediction approach.[11,13]

5. Host–Pathogen Interaction Data Collection and Integration

The rapid progress on the host–pathogen interaction studies is supported by many collection, dissemination, integration, analysis and visualization tools. Host–pathogen interaction databases, can be divided into two categories: (i) collection and curation databases and (ii) integration and analysis databases. There is no clear dividing line for the two categories. This categorization is mostly for convenience of discussion.

5.1. *Host–Pathogen Interaction Data Collection Techniques*

Text mining is frequently used for extracting PPI data from literature. This is very useful in facilitating the manual curation of host–pathogen interaction data from publications. For example, VirusMINT[16] relies on a simple text mining approach based on a context-free grammar that identifies sentences containing interaction information to select relevant articles. VirHostNet[17] also uses a text mining approach to prioritize papers for manual curation, where the text mining pipeline is applied to extract keywords related to both virus and experimental procedures.

Moreover, text mining techniques have been applied to specifically extract host–pathogen PPIs from biomedical literature with considerable accuracy.[71] Feature-based and language-based approaches are introduced and compared by Thieu *et al.*[71] Both methods can automatically detect host–pathogen interaction data and extract information about organisms and proteins involved in the interactions.[71] The feature-based method uses SVM trained on features derived from the individual sentences, including names of the organisms and corresponding proteins or genes, keywords describing host–pathogen interaction-specific information, general PPI information, experimental methods and other statistical information.[71]

The language-based method uses a link grammar parser combined with semantic patterns derived from training examples.[71]

5.2. *Host–Pathogen Interaction Collection and Curation Databases*

Host–pathogen interaction collection and curation databases are those dedicated to collect and curate host–pathogen interaction from literature or from experimental data. These databases may have imported some parts of data from other databases but at least contain some data derived from their own collection or curation. Collection and curation databases serve primarily as "data source," and generally provide only simple tools for searching, visualization or analysis. They are often used as the data source for host–pathogen interaction studies or are imported by integration and analysis databases (to be discussed in the next section). In this section we list some representative databases of this category.

PHI-base is a database created to catalog experimentally verified pathogenicity, virulence and effector genes of fungal and Oomycete pathogens.[72] After its update, the PHI-base also covers bacterial pathogens. The pathogens covered by PHI-base infect a wide range of hosts.[18]

The "HIV-1, Human Protein Interaction database" at NCBI aims at cataloging all interactions between HIV-1 and human proteins published in the peer-reviewed literature.[19] Basic search and visualization tools are also provided. It is very popular among the AIDS research community and is well known for its intensive long-term curation effort. The *H. sapiens*–HIV-1 interaction data included in this database cover both direct and indirect interactions; brief description and PubMed IDs are also provided for each entry. Its *H. sapiens*–HIV-1 PPI data have been used in several studies[10,11,13] and imported as source data by some databases.[16]

The VirusMINT database aims at collecting all interactions between viral and human proteins reported in the literature.[16] It covers more than 110 different viral strains.[16] The curation effort has focused mainly on viruses known to be associated with infectious diseases and oncogenesis in humans.[16] VirusMINT derives its host–virus PPI data from two sources. The first source is from databases of literature-curated PPIs like IntAct,[33] MINT[30] and "HIV-1, Human Protein Interaction database".[19] Host-virus PPI data are uploaded from IntAct and MINT directly without further curation. Only a subset of "HIV-1, Human Protein Interaction database" is imported, which pertains to enzymatic reactions, physical associations and co-localization. The second source is manually curated PPIs from literature; the PPIs are first uploaded to MINT and then re-imported into VirusMINT.[16] The literature curation is facilitated by simple text ming techniques in selecting relevant articles. MINT[30] and VirusMINT are both curated by MINT curators and uploaded first to MINT then to VirusMINT. Much of the PPI data in VirusMINT are the same as in MINT. VirusMINT also provides searching and visualization functions.

VirHostNet (Virus–Host Network) is a management and analysis database of integrated virus–virus, virus–host and host–host interaction networks and their functional annotations.[17] The interaction data are reconstructed from public databases and, for virus–virus and virus–host interactions, are also supplemented by original literature-curated dataset.

A simple text mining strategy has been adopted for prioritizing articles for literature curation. Virus–virus and virus–host interactions data from public databases are also carefully inspected before importing into VirHostNet.[17] Search and visualization functions are supported in this database.

The databases mentioned below are mostly well known for their intra-species PPI datasets. However, their curation and collection have also been extended to inter-species host–pathogen PPIs. IntAct is an open-source, open-data molecular interaction database.[73] Both intra- and inter-species PPI data are collected in this database either from the literature or from direct data depositions. For each PPI entry, a brief description, experimental method and literature citation are included. Several integration databases[16,21,23] import host–pathogen PPI data from IntAct. It is well known for its intensive curation and quality control process. BioGRID (Biological General Repository for Interaction Datasets) archives and disseminates genetic and protein interaction data.[74] BioGRID interaction data are curated from both high-throughput experiments and individual focused studies. Most of the interaction data are intra-species PPIs, but some host–pathogen PPIs are included. DIP (Database of Interacting Proteins) aims to integrate the diverse experimental evidences on PPIs into the database.[28] It is another well-known intra-species PPI integration database. It also collects host–pathogen PPI data. Reactome is a curated, peer-reviewed knowledgebase of biological pathways.[32] It curates both intra- and inter-species data. Curated host–pathogen PPI data are also available in Reactome.[32] BIND (Biomolecular interaction network database) archives biomolecular interaction, complex and pathway information, and is a major source of curated biomolecular interactions.[47] It has not been maintained for the last few years, until a recent update and conversion of the BIND data to a standard format (Proteomics Standard Initiative-Molecular Interaction 2.5).[75] Its main interaction data are intra-species PPIs, but also contains some host–pathogen PPI data.

5.3. *Host–Pathogen Interaction Integration and Analysis Databases*

Host–pathogen interaction integration and analysis databases mainly integrate host–pathogen interaction data from other source databases. While they usually do not have their own intensive curation process, some of them provide powerful analysis and visualization functions. The integrated data can be more than just host–pathogen PPI data, like gene expression data related to infection, disease outbreak information, pathogen proteomics data, protein functional data, protein complex data, etc. In this section, representative integration and analysis databases are briefly introduced.

APID (Agile Protein Interaction Data Analyzer) provides an open-access framework where all known experimentally validated PPIs (BIND, BioGRID, DIP, HPRD, IntAct and MINT) are unified in it.[76] iRefIndex[77] provides an index of PPIs from BIND, BioGrid, DIP, HPRD, IntAct, MINT, MPact,[78] MIPS[50] and OPHID.[79] iRefWeb[80] provides a searchable web interface to the iRefIndex. Both APID and iRefIndex (iRefWeb) are general PPI integration databases, unlike the following databases which are dedicated to host–pathogen interaction data integration and analysis. They include host–pathogen PPIs just because their source databases contain some host–pathogen PPI data.

PHIDIAS (Pathogen–Host Interaction Data Integration and Analysis System) includes six components (PGBrowser, Pacodom, BLAST searches, Phinfo, Phigen and Phinet) for searching, comparing and analyzing integrated genome sequences, conserved domains, host–pathogen interaction data and gene expression data related to host–pathogen interactions.[20]

HPIDB is a host–pathogen PPI database which integrates experimental PPIs from several public databases (BIND, REACTOME, MINT, IntAct, PIG).[21] Some of the HPIDB sources may have content overlap with each other, since PIG[23] also integrates data from BIND, REACTOME and

MINT. Different from PIG — which only considers one host, *H. sapiens* — HPIDB also takes other hosts into account.

GPS-Prot is an integration and visualization database that currently focuses on *H. sapiens*–HIV interactions.[22] It allows for integration of different HIV interaction data types.[22] Human PPI data are imported from the following six databases, MINT, IntAct, DIP, MIPS, BioGRID and HPRD. *H. sapiens*–HIV PPI data are import from VirusMINT.[16] The GPS-Prot can group proteins into functional modules or protein complexes, generating intuitive network representations. It allows for the uploading of user-generated data.[22]

RCBPR (Resource Center for Biodefense Proteomics Research) is a bioinformatics framework employing a protein-centric approach to integrate and the collect large and heterogeneous data.[81] It is no longer functional and the collected data have been transferred to the Pathogen Portal (http://www.pathogenportal.org).

PIG (Pathogen Interaction Gateway)[23] is created by integrating host–pathogen PPI data from a number of public resources, including BIND, REACTOME, MINT, MIPS, HPRD, DIP and MvirDB.[82] Now PIG has become part of the PATRIC[25] database but only the bacterial pathogen data in PIG have been merged into PATRIC (which primarily focuses on bacterial pathogens) and can still be accessed at http://www.patricbrc.org/portal/portal/patric/HPITool.

Disease View is a host–pathogen data integration and visualization resource that enables access, analysis and integration of diverse data sources, including host, pathogen, host–pathogen interactions and disease outbreak. It provides a mechanism for infectious-disease–centric data analysis and visualization. The infectious diseases covered by Disease View come with related information like the corresponding pathogen that causes the infectious diseases, the associated pathogen virulence genes and the genetic and chemical evidences for the human genes that are associated with the diseases.[24] It is implemented as a component of PATRIC.[25]

PATRIC (the Pathosystems Resource Integration Center) is a comprehensive genomics-centric relational database for infectious-disease research.[25] Comprehensive bacterial genomics data, associated data relevant to genomic analysis and analysis tools and platforms have been provided in this database. Its resources can be divided into two categories, (i) organisms, genomes and comparative genomics and (ii) recurrent integration of community-derived associated data.

5.4. *Host–Pathogen Interaction Integration and Analysis Software*

Not only databases but also standalone software tools are available for host–pathogen interaction studies.

Conventional complex network analysis and visualization software platforms like Cytoscape[83] continue to be very popular in host–pathogen interaction studies. Cytoscape has been used for visualization of host–pathogen PPI networks in several works.[9,49] Software that are specifically designed for host–pathogen interaction studies have also been developed. For example, BiologicalNetworks is a system that enables the integration of multiscale data for host–pathogen studies.[26] It can integrate diverse experimental data types, including molecular interactions, phylogenetic classifications, genomic sequences, protein structure information, gene expression, pathway and virulence data for host–pathogen studies.[26] It provides several useful functions, including analyzing subnetworks, building host–pathogen interaction networks, studying individual genes, identifying potential drug targets, adding phylogeography, integrating user data, etc.[26] This system is available through a standalone Java application (BiologicalNetworks), which provides complex data analysis capabilities, and a web interface (http://flu.sdsc. edu) for quick search of phylogenetic relations among sequenced strains.

6. Discussion

6.1. *Contributions and Limitations of Current Host–Pathogen Interaction Study Approaches*

The current host–pathogen interaction studies described in this work are indispensable stepping stones for the future progress in this field. Nevertheless, several limitations are also noticeable.

6.1.1. *Contributions of Current Host–Pathogen Interaction Studies*

Usually host–pathogen PPIs prediction followed by analyses and assessment would produce enriched datasets which are useful for the experimental testing and verification. This could save a lot of wet lab experimental effort. The prediction and verification approaches discussed in these pioneering works pave the way for future development of host–pathogen interaction studies as they provide insights for improvements and basis for comparison.

6.1.2. *Limitations of Current Host–Pathogen Interaction Prediction Approaches*

It is not uncommon that different prediction approaches yield very different prediction results, even in the same host–pathogen system, as revealed by the comparison among different *H. sapiens*–HIV PPI datasets generated from different prediction approaches.[8]

It has not escaped our notice that some publications repeatedly report almost the same prediction method whose performance and predicted results have not been rigorously assessed. Sometimes even the source data (like template PPI data) are the same, yet only applied to different host–pathogen systems.[2–4] Therefore the contribution of these publications may be relatively limited.

Limited by the current understanding of host–pathogen protein interaction, the prediction approaches may not resemble the real biological scenario. For example, although the approach based on motif–domain interaction[10] has achieved good performance, Evans *et al.* have also mentioned the mismatches between predicted result and gold standard may be caused by the fact that real mechanisms of host–pathogen PPIs are more complicated than the assumption (that host–pathogen PPIs are mediated by ELMs-CDs interactions) in this study.

6.1.3. *Limitations of Current Host–Pathogen Interaction Verification Approaches*

Due to the limitation of current known gold-standard host–pathogen PPI data and limited understanding of the host–pathogen interactions, most current assessments are rather "indirect" approaches.

Some verifications may not have a strong logical or biological basis. For example, Dyer *et al.*[9] assessed predicted *H. sapiens–P. falciparum* PPIs by examining whether the pairs of human proteins predicted to interact with the same pathogen proteins are close to each other in the human PPI network. This assessment through distance in triplets may not have biological or experimental basis. However, based on the observed topological properties discussed in the Sec. 3, calculating whether the human proteins targeted by predicted PPIs have shorter paths to other reachable proteins in the human interactome, would serve as a possible assessment. Dyer *et al.*[9] also analyze the gene expression profile of pathogen protein pairs interacting with the same host proteins; they report that those pathogen protein pairs exhibit correlated gene expression profile, and also the same for host protein pairs interacting with same pathogen proteins. While gene expression profile can be reasonably used in assessing *M. tuberculosis* H37Rv intra-species PPI datasets as done by Zhou and Wong,[84] it may lack biological basis in assessing inter-species host–pathogen PPI dataset through gene expression in the form of triplets as conducted by Dyer *et al.*[9]

Explanation on selected examples of predicted results reflects neither the quality of the whole predicted results nor the performance of prediction approaches. For example, biological explanation for selected examples should not be used as the only assessment of a few predicted results, as what we observed in several studies[1–4] — the qualities of those prediction results are still largely in doubt.

6.2. *Contributions and Limitations of Current Host–Pathogen Interaction Databases*

Current host–pathogen interaction databases contribute a lot to host–pathogen interaction studies in the form of collecting and integrating valuable host–pathogen interactions and providing powerful analysis tools. Yet some possible limitations also exist.

The host–pathogen interaction databases greatly facilitate host–pathogen interaction studies in collecting and integrating valuable interaction and related genomic and experimental data scattered in primary literature. Without these databases, many of the studies described in this review would be impossible or at least would take much longer time and more effort in collecting the source data. Moreover, these databases provide the platforms for accessing and sharing of host–pathogen interaction data, which in turn facilitate research in this field. Many databases not only enable convenient data access and integration of related host and pathogen data but also provide powerful analysis tools which significantly increase the efficiency of host–pathogen interactions analysis.

Some databases lack long-term support and are no longer in function, like RCBPR.[81,85] And there are some information loss in the merging of the one database into another, like PIG,[23] where only its bacterial pathogen data have been moved into PATRIC. Some databases, although still in operation, lack necessary updates, like ViursMINT[16] and HPIDB.[21]

6.3. *Literature-curated Host–Pathogen Interaction Data*

The literature-curated interaction data from the databases discussed above are often used as gold standard in studies on host–pathogen interactions. However, a study[86] on intra-species PPI datasets shows that literature-curated PPI data may not be as accurate as people usually have assumed. Therefore, those manually curated host–pathogen PPI data should be used with caution. For example, the "HIV-1, Human Protein Interaction database" at NCBI[19] has been divided into "positively labeled" and "partially labeled" data in Qi *et al.*,[13] and VirusMINT only imports a portion of the PPI data from it.

6.4. *Future Development of Host–Pathogen Interaction Studies*

Fundamental progress in host–pathogen interaction studies will be achieved in the future, due to better source data and improved investigation approaches and tools, and these will lead to deeper understanding of host–pathogen interaction.

It is reasonable to expect that high-quality source data will become increasingly available. More genomic and proteomic data will come out. As a result, more accurate orthologs can be identified between less-known pathogens and well-studied organisms. This will enable the application of homology-based approach to many understudied host–pathogen systems. Better annotation of known motifs and counter domains will result in enhanced performance of domain–motif interaction-based prediction approaches.[10] With more protein structures being resolved, the structure-based approach will have higher-quality structural template and with much larger coverage. With more high-resolution Structural Interaction Network (SIN) being provided to analysis, more fundamental interaction mechanisms will come to light. Abundant and accurate functional information — including GO annotation, gene expression, RNA interference and pathway data — will largely improve the performance of current analysis approaches. More reliable PPI data (both intra- and inter-species) will provide sufficient high-quality templates for homology-based approaches and also larger as well as more accurate training and testing data for machine-learning–based approaches. The lack of gold-standard host–pathogen PPI data will also be alleviated in the future. As a result more direct and effective verification approaches will be available for many host–pathogen systems. With better source data from a variety of aspects, the prediction approaches that can integrate different types of data (e.g. machine learning-based approach) into their prediction will have good potential.

More effective host–pathogen interaction prediction algorithms will be proposed in the future. For example, the core algorithm used by Dyer *et al.* is an association method proposed in 2001. However, several other algorithms with enhanced performance are available now, including the association numerical method (ASNM)[87] in 2004 and the association probabilistic method (APM)[88] in 2006. Using ASNM or APM in predicting host–pathogen PPIs may improve prediction performance. Recently, Itzhaki *et al.*[37] propose the concept of "preferential use of protein domain pairs as interaction mediators" may also introduce new idea to DDI-based prediction algorithms. More accurate prediction and more effective verification approaches on a better understanding of host–pathogen interactions will come out.

All these will help the community to achieve the ultimate goal of better prevention and treatment of infectious diseases.

Acknowledgments

We thank Sriganesh Srihari for the critical reading of this paper. This project was supported in part by an NGS scholarship and a Singapore Ministry of Education Tier-2 grant MOE2009-T2-2-004.

References

1. Lee SA, Chan C, Tsai CH, Lai JM, Wang FS, Kao CY, Huang CY, Ortholog-based protein–protein interaction prediction and its application to inter-species interactions, *BMC Bioinformatics* **9**(Suppl 12):S11, 2008.
2. Krishnadev O, Srinivasan N, A data integration approach to predict host–pathogen protein–protein interactions: Application to recognize protein interactions between human and a malarial parasite, *In Silico Biol* **8**(3):235–250, 2008.
3. Tyagi N, Krishnadev O, Srinivasan N, Prediction of protein–protein interactions between *Helicobacter pylori* and a human host, *Mol BioSyst* **5**(12):1630–1635, 2009.
4. Krishnadev O, Srinivasan N, Prediction of protein–protein interactions between human host and a pathogen and its application to three pathogenic bacteria, *Int J Biol Macromol* **48**:613–619, 2011.

5. Wuchty S, Computational prediction of host-parasite protein interactions between *P. falciparum* and *H. sapiens*, *PLoS ONE* **6**(11):e26960, 2011.

6. Davis FP, Barkan DT, Eswar N, McKerrow JH, Sali A, Host–pathogen protein interactions predicted by comparative modeling, *Protein Sci* **16**(12):2585–2596, 2007.

7. Doolittle JM, Gomez SM, Mapping protein interactions between Dengue virus and its human and insect hosts, *PLoS Negl Trop Disease* **5**(2):e954, 2011.

8. Janet D, Shawn G, Structural similarity-based predictions of protein interactions between HIV-1 and *Homo sapiens*, *Virol J* **7**:82, 2010.

9. Dyer MD, Murali TM, Sobral BW, Computational prediction of host–pathogen protein–protein interactions, *Bioinformatics* **23**(13):i159–i166, 2007.

10. Evans P, Dampier W, Ungar L, Tozeren A, Prediction of HIV-1 virus-host protein interactions using virus and host sequence motifs, *BMC Med Genomics* **2**(1):27, 2009.

11. Tastan O, Qi Y, Carbonell JG, Klein-Seetharaman J, Prediction of interactions between HIV-1 and human proteins by information integration, *Pacific Symp Biocomput* **14**:516–527, 2009.

12. Dyer MD, Murali TM, Sobral BW, Supervised learning and prediction of physical interactions between human and HIV proteins, *Infect Genet Evol* **11**(5):917–923, 2011.

13. Qi Y, Tastan O, Carbonell JG, Klein-Seetharaman J, Weston J, Semi-supervised multi-task learning for predicting interactions between HIV-1 and human proteins, *Bioinformatics* **26**(18):i645–i652, 2010.

14. Singh I, Tastan O, Klein-Seetharaman J, Comparison of virus interactions with human signal transduction pathways, *Proc First ACM Int Conf Bioinform Comput Biol*, pp. 17–24, 2010.

15. Zhao Z, Xia J, Tastan O, Singh I, Kshirsagar M, Carbonell J, Klein-Seetharaman J, Virus interactions with human signal transduction pathways, *Int J Comput Biol Drug Design* **4**(1):83–105, 2011.

16. Chatr-aryamontri A, Ceol A, Peluso D, Nardozza A, Panni S, Sacco F, Tinti M, Smolyar A, Castagnoli L, Vidal M *et al.*, VirusMINT: A viral protein interaction database, *Nucleic Acids Res* **37**(Suppl 1):D669–D673, 2009.

17. Navratil V, De Chassey B, Meyniel L, Delmotte S, Gautier C, André P, Lotteau V, Rabourdin-Combe C, VirHostNet: A knowledge base for the management and the analysis of proteome-wide virus–host interaction networks, *Nucleic Acids Res* **37**(Suppl 1):D661–D668, 2009.

18. Winnenburg R, Urban M, Beacham A, Baldwin T, Holland S, Lindeberg M, Hansen H, Rawlings C, Hammond-Kosack K, Köhler J, PHI-base update: Additions to the pathogen–host interaction database, *Nucleic Acids Res* **36**(Suppl 1):D572–D576, 2008.

19. Fu W, Sanders-Beer B, Katz K, Maglott D, Pruitt K, Ptak R, Human immunodeficiency virus type 1, human protein interaction database at NCBI, *Nucleic Acids Res* **37**(Suppl 1):D417–D422, 2009.

20. Xiang Z, Tian Y, He Y *et al.*, PHIDIAS: A pathogen-host interaction data integration and analysis system, *Genome Biol* **8**(7):R150, 2007.

21. Ranjit K, Bindu N, HPIDB-a unified resource for host–pathogen interactions, *BMC Bioinformatics* **11**(Suppl 6):S16, 2010.

22. Fahey M, Bennett M, Mahon C, Jäger S, Pache L, Kumar D, Shapiro A, Rao K, Chanda S, Craik C *et al.*, GPS-Prot: A web-based visualization platform for integrating host–pathogen interaction data, *BMC Bioinformatics* **12**(1):298, 2011.

23. Driscoll T, Dyer M, Murali T, Sobral B, PIG — the pathogen interaction gateway, *Nucleic Acids Res* **37**(Suppl 1):D647–D650, 2009.

24. Driscoll T, Gabbard J, Mao C, Dalay O, Shukla M, Freifeld C, Hoen A, Brownstein J, Sobral B, Integration and visualization of host–pathogen data related to infectious diseases, *Bioinformatics* **27**(16):2279–2287, 2011.

25. Gillespie J, Wattam A, Cammer S, Gabbard J, Shukla M, Dalay O, Driscoll T, Hix D, Mane S, Mao C *et al.*, PATRIC: The comprehensive bacterial bioinformatics resource with a focus on human pathogenic species, *Infect Immun* **79**(11):4286–4298, 2011.

26. Sergey K, Mayya S, Yulia D, Amarnath G, Animesh R, Julia P, Michael B, BiologicalNetworks-tools enabling the integration of multi-scale data for the host–pathogen studies, *BMC Syst Biol* **5**:7, 2011.

27. Matthews L, Vaglio P, Reboul J, Ge H, Davis B, Garrels J, Vincent S, Vidal M, Identification of potential interaction networks using sequence-based searches for conserved protein–protein interactions or interologs, *Genome Res* **11**(12):2120–2126, 2001.

28. Salwinski L, Miller C, Smith A, Pettit F, Bowie J, Eisenberg D, The database of interacting proteins: 2004 update, *Nucleic Acids Res* **32**(Suppl 1):D449–D451, 2004.

29. Finn R, Marshall M, Bateman A, iPfam: Visualization of protein–protein interactions in PDB at domain and amino acid resolutions, *Bioinformatics* **21**(3):410–412, 2005.

30. Zanzoni A, Montecchi-Palazzi L, Quondam M, Ausiello G, Helmer-Citterich M, Cesareni G, MINT: A Molecular INTeraction database, *FEBS Lett* **513**(1):135–140, 2002.

31. Mishra G, Suresh M, Kumaran K, Kannabiran N, Suresh S, Bala P, Shivakumar K, Anuradha N, Reddy R, Raghavan T *et al.*, Human protein reference database — 2006 update, *Nucleic Acids Res* **34**(Suppl 1):D411–D414, 2006.

32. Joshi-Tope G, Gillespie M, Vastrik I, D'Eustachio P, Schmidt E, Bono Bd, Jassal B, Gopinath G, Wu G, Matthews L *et al.*, Reactome: A

knowledgebase of biological pathways, *Nucleic Acids Res* **33**(Suppl 1): D428–D432, 2005.

33. Hermjakob H, Montecchi-Palazzi L, Lewington C, Mudali S, Kerrien S, Orchard S, Vingron M, Roechert B, Roepstorff P, Valencia A *et al.*, IntAct: An open source molecular interaction database, *Nucleic Acids Res* **32**(Suppl 1):D452–D455, 2004.

34. Murzin A, Brenner S, Hubbard T, Chothia C *et al.*, Scop: A structural classification of proteins database for the investigation of sequences and structures, *J Mol Biol* **247**(4):536–540, 1995.

35. Davis F, Sali A, PIBASE: A comprehensive database of structurally defined protein interfaces, *Bioinformatics* **21**(9):1901–1907, 2005.

36. Holm L, Kääriäinen S, Rosenström P, Schenkel A, Searching protein structure databases with DaliLite v. 3, *Bioinformatics* **24**(23):2780–2781, 2008.

37. Itzhaki Z, Akiva E, Margalit H, Preferential use of protein domain pairs as interaction mediators: Order and transitivity, *Bioinformatics* **26**(20):2564–2570, 2010.

38. Sprinzak E, Margalit H, Correlated sequence-signatures as markers of protein–protein interaction, *J Mol Biol* **311**(4):681–692, 2001.

39. Quevillon E, Silventoinen V, Pillai S, Harte N, Mulder N, Apweiler R, Lopez R, InterProScan: Protein domains identifier, *Nucleic Acids Res* **33**(Suppl 2):W116–W120, 2005.

40. Kim W, Kim K, Lee E, Marcotte E, Kim H, Suh J, Identification of disease specific protein interactions between the gastric cancer causing pathogen, *H. pylori*, and human hosts using protein network modeling and gene chip analysis, *BioChip J* **1**:179–187, 2007.

41. Kim W, Park J, Suh J *et al.*, Large scale statistical prediction of protein–protein interaction by potentially interacting domain (PID) pair, *Genome Inform Ser* **13**:42–50, 2002.

42. Han D, Kim H, Jang W, Lee S, Suh J, PreSPI: A domain combination based prediction system for protein–protein interaction, *Nucleic Acids Res* **32**(21):6312–6320, 2004.

43. Edwards R, Davey N, Shields D, SLiMFinder: A probabilistic method for identifying over-represented, convergently evolved, short linear motifs in proteins, *PLoS ONE* **2**(10):e967, 2007.

44. Hugo W, Ng S, Sung W, D-SLIMMER: Domain-SLiM interaction motifs miner for sequence based protein–protein interaction data, *J Proteome Res* **10**(12):5285–5295, 2011.

45. Puntervoll P, Linding R, Gemünd C, Chabanis-Davidson S, Mattingsdal M, Cameron S, Martin D, Ausiello G, Brannetti B, Costantini A *et al.*, ELM server: A new resource for investigating short functional sites in modular eukaryotic proteins, *Nucleic Acids Res* **31**(13):3625–3630, 2003.

46. Hulo N, Bairoch A, Bulliard V, Cerutti L, Cuche B, De Castro E, Lachaize C, Langendijk-Genevaux P, Sigrist C, The 20 years of PROSITE, *Nucleic Acids Res* **36**(Suppl 1):D245–D249, 2008.

47. Gilbert D, Biomolecular interaction network database, *Briefings Bioinform* **6**(2):194–198, 2005.

48. Calderwood M, Venkatesan K, Xing L, Chase M, Vazquez A, Holthaus A, Ewence A, Li N, Hirozane-Kishikawa T, Hill D *et al.*, Epstein–Barr virus and virus human protein interaction maps, *Proc Nat Acad Sci USA* **104**(18):7606–7611, 2007.

49. Dyer M, Murali T, Sobral B, The landscape of human proteins interacting with viruses and other pathogens, *PLoS Pathogen* **4**(2):e32, 2008.

50. Pagel P, Kovac S, Oesterheld M, Brauner B, Dunger-Kaltenbach I, Frishman G, Montrone C, Mark P, Stümpflen V, Mewes H *et al.*, The MIPS mammalian protein–protein interaction database, *Bioinformatics* **21**(6):832–834, 2005.

51. Dyer M, Neff C, Dufford M, Rivera C, Shattuck D, Bassaganya-Riera J, Murali T, Sobral B, The human-bacterial pathogen protein interaction networks of *Bacillus anthracis, Francisella tularensis* and *Yersinia pestis, PLoS ONE* **5**(8):e12089, 2010.

52. Franzosa E, Xia Y, Structural principles within the human-virus protein–protein interaction network, *Proc Nat Acad Sci USA* **108**(26):10538–10543, 2011.

53. Rappoport N, Linial M, Viral proteins acquired from a host converge to simplified domain architectures, *PLoS Comput Biol* **8**(2):e1002364, 2012.

54. Beißbarth T, Speed T, Gostat: Find statistically overrepresented gene ontologies within a group of genes, *Bioinformatics* **20**(9):1464–1465, 2004.

55. Boyle E, Weng S, Gollub J, Jin H, Botstein D, Cherry J, Sherlock G, GO:: TermFinder open source software for accessing gene ontology information and finding significantly enriched gene ontology terms associated with a list of genes, *Bioinformatics* **20**(18):3710–3715, 2004.

56. Bauer S, Grossmann S, Vingron M, Robinson P, Ontologizer 2.0 a multifunctional tool for go term enrichment analysis and data exploration, *Bioinformatics* **24**(14):1650–1651, 2008.

57. Dennis Jr G, Sherman B, Hosack D, Yang J, Gao W, Lane H, Lempicki R *et al.*, DAVID: Database for annotation, visualization, and integrated discovery, *Genome Biol* **4**(5):P3, 2003.

58. Balakrishnan S, Tastan O, Carbonell J, Klein-Seetharaman J, Alternative paths in HIV-1 targeted human signal transduction pathways, *BMC Genom* **10**(Suppl 3):S30, 2009.

59. Schaefer C, Anthony K, Krupa S, Buchoff J, Day M, Hannay T, Buetow K, PID: The pathway interaction database, *Nucleic Acids Res* **37**(Suppl 1): D674–D679, 2009.

60. Sassetti C, Rubin E, Genetic requirements for mycobacterial survival during infection, *Proc Nat Acad Sci USA* **100**(22):12989–12994, 2003.

61. Rachman H, Strong M, Ulrichs T, Grode L, Schuchhardt J, Mollenkopf H, Kosmiadi G, Eisenberg D, Kaufmann S, Unique transcriptome signature of *Mycobacterium tuberculosis* in pulmonary tuberculosis, *Infect Immun* **74**(2):1233–1242, 2006.

62. Chaussabel D, Semnani R, McDowell M, Sacks D, Sher A, Nutman T, Unique gene expression profiles of human macrophages and dendritic cells to phylogenetically distinct parasites, *Blood* **102**(2):672–681, 2003.

63. König R, Zhou Y, Elleder D, Diamond T, Bonamy G, Irelan J, Chiang C, Tu B, De Jesus P, Lilley C *et al.*, Global analysis of host–pathogen interactions that regulate early-stage HIV-1 replication, *Cell* **135**(1):49–60, 2008.

64. Brass A, Dykxhoorn D, Benita Y, Yan N, Engelman A, Xavier R, Lieberman J, Elledge S, Identification of host proteins required for HIV infection through a functional genomic screen, *Science* **319**(5865):921–926, 2008.

65. Zhou H, Xu M, Huang Q, Gates A, Zhang X, Castle J, Stec E, Ferrer M, Strulovici B, Hazuda D *et al.*, Genome-scale RNAi screen for host factors required for HIV replication, *Cell Host Microbe* **4**(5):495–504, 2008.

66. Yeung M, Houzet L, Yedavalli V, Jeang K, A genome-wide short hairpin RNA screening of jurkat T-cells for human proteins contributing to productive HIV-1 replication, *J Biol Chem* **284**(29):19463–19473, 2009.

67. Sessions O, Barrows N, Souza-Neto J, Robinson T, Hershey C, Rodgers M, Ramirez J, Dimopoulos G, Yang P, Pearson J *et al.*, Discovery of insect and human dengue virus host factors, *Nature* **458**(7241):1047–1050, 2009.

68. Krishnan M, Ng A, Sukumaran B, Gilfoy F, Uchil P, Sultana H, Brass A, Adametz R, Tsui M, Qian F *et al.*, RNA interference screen for human genes associated with West Nile virus infection, *Nature* **455**(7210):242–245, 2008.

69. Chertova E, Chertov O, Coren L, Roser J, Trubey C, Bess Jr J, Sowder II R, Barsov E, Hood B, Fisher R *et al.*, Proteomic and biochemical analysis of purified human immunodeficiency virus type 1 produced from infected monocyte-derived macrophages, *J Virol* **80**(18):9039–9052, 2006.

70. Ott D, Cellular proteins detected in HIV-1, *Rev Med Virol* **18**(3):159–175, 2008.

71. Thieu T, Joshi S, Warren S, Korkin D, Literature mining of host–pathogen interactions: Comparing feature-based supervised learning and language-based approaches, *Bioinformatics* **28**(6):867–875, 2012.

72. Winnenburg R, Baldwin T, Urban M, Rawlings C, Köhler J, Hammond-Kosack K, PHI-base: A new database for pathogen host interactions, *Nucleic Acids Res* **34**(Suppl 1):D459–D464, 2006.

73. Kerrien S, Aranda B, Breuza L, Bridge A, Broackes-Carter F, Chen C, Duesbury M, Dumousseau M, Feuermann M, Hinz U *et al.*, The IntAct

molecular interaction database in 2012, *Nucleic Acids Res* **40**(D1):D841–D846, 2012.

74. Stark C, Breitkreutz B, Chatr-Aryamontri A, Boucher L, Oughtred R, Livstone M, Nixon J, Van Auken K, Wang X, Shi X *et al.*, The BioGRID interaction database: 2011 update, *Nucleic Acids Res* **39**(Suppl 1):D698–D704, 2011.

75. Isserlin R, El-Badrawi R, Bader G, The biomolecular interaction network database in PSI-MI 2.5, *Database: J Biol Databases Curation* **2011**, 2011.

76. Prieto C, De Las Rivas J, APID: Agile Protein Interaction Dataanalyzer, *Nucleic Acids Res* **34**(Suppl 2):W298–W302, 2006.

77. Razick S, Magklaras G, Donaldson I, iRefIndex: A consolidated protein interaction database with provenance, *BMC Bioinformatics* **9**(1):405, 2008.

78. Güldener U, Münsterkötter M, Oesterheld M, Pagel P, Ruepp A, Mewes H, Stümpflen V, MPact: The MIPS protein interaction resource on yeast, *Nucleic Acids Res* **34**(Suppl 1):D436–D441, 2006.

79. Brown K, Jurisica I, Online predicted human interaction database, *Bioinformatics* **21**(9):2076–2082, 2005.

80. Turner B, Razick S, Turinsky A, Vlasblom J, Crowdy E, Cho E, Morrison K, Donaldson I, Wodak S, iRefWeb: Interactive analysis of consolidated protein interaction data and their supporting evidence, *Database: J Biol Databases Curation* **2010**, 2010.

81. McGarvey P, Huang H, Mazumder R, Zhang J, Chen Y, Zhang C, Cammer S, Will R, Odle M, Sobral B *et al.*, Systems integration of biodefense omics data for analysis of pathogen-host interactions and identification of potential targets, *PLoS ONE* **4**(9):e7162, 2009.

82. Zhou C, Smith J, Lam M, Zemla A, Dyer M, Slezak T, MvirDB a microbial database of protein toxins, virulence factors and antibiotic resistance genes for bio-defence applications, *Nucleic Acids Res* **35**(Suppl 1):D391–D394, 2007.

83. Smoot M, Ono K, Ruscheinski J, Wang P, Ideker T, Cytoscape 2.8: New features for data integration and network visualization, *Bioinformatics* **27**(3):431–432, 2011.

84. Zhou H, Wong L, Comparative analysis and assessment of *M. tuberculosis* H37Rv protein–protein interaction datasets, *BMC Genomics* **12**(Suppl 3): S20, 2011.

85. Zhang C, Crasta O, Cammer S, Will R, Kenyon R, Sullivan D, Yu Q, Sun W, Jha R, Liu D *et al.*, An emerging cyberinfrastructure for biodefense pathogen and pathogen–host data, *Nucleic Acids Res* **36**(Suppl 1):D884–D891, 2008.

86. Cusick M, Yu H, Smolyar A, Venkatesan K, Carvunis A, Simonis N, Rual J, Borick H, Braun P, Dreze M *et al.*, Literature-curated protein interaction datasets, *Nature Meth* **6**(1):39–46, 2008.

87. Hayashida M, Ueda N, Akutsu T *et al.*, A simple method for inferring strengths of protein–protein interactions, *Genome Inform Ser* **15**(1):56–68, 2004.
88. Chen L, Wu L, Wang Y, Zhang X, Inferring protein interactions from experimental data by association probabilistic method, *Proteins: Struct Funct Bioinform* **62**(4):833–837, 2006.

Hufeng Zhou received both his Bachelor of Arts degree from Huazhong University of Science and Technology and Bachelor of Engineering degree from Huazhong Agricultural University in Wuhan, Hubei, P. R. China, in 2009. Hufeng is currently a Ph.D. candidate at the National University of Singapore. He mainly focuses on the research topics related to host–pathogen interactions, intra-/inter-species protein–protein interaction, pathways integration and analysis, etc.

Jingjing Jin received her Bachelor degree from Sichuan University in Chengdu, Sichuan, China, in 2009. Jingjing is currently a Ph.D. candidate at the National University of Singapore. She mainly focuses on the research topics related to long noncoding RNA and next generation sequencing, etc.

Limsoon Wong is a Professor in the School of Computing and the School of Medicine at the National University of Singapore. He is currently working mostly on knowledge discovery technologies and is especially interested in their application to biomedicine. He serves on the editorial boards of the *Journal of Bioinformatics and Computational Biology* (ICP), *Bioinformatics* (OUP), and *Drug Discovery Today* (Elsevier).

CHAPTER 6

Training Host-Pathogen Protein–Protein Interaction Predictors[a]

Abdul Hannan Basit[*,†,‡], Wajid Arshad Abbasi[*,§], Amina Asif[*,¶],
Sadaf Gull[*,‖] and Fayyaz Ul Amir Afsar Minhas[*,**]

*Department of Computer and Information Sciences
Biomedical Informatics Research Laboratory
Pakistan Institute of Engineering and Applied Sciences (PIEAS)
Nilore, Islamabad 44000, Pakistan*

†*Department of Electrical Engineering
Pakistan Institute of Engineering and Applied Sciences (PIEAS)
Nilore, Islamabad 44000, Pakistan*
‡*hannan.ahb@gmail.com*
§*wajidarshad@gmail.com*
¶*a.asif.shah01@gmail.com*
‖*sadafzakarkhan@gmail.com*
**afsar@pieas.edu.pk*

Detection of protein–protein interactions (PPIs) plays a vital role in molecular biology. Particularly, pathogenic infections are caused by interactions of host and pathogen proteins. It is important to identify host–pathogen interactions (HPIs) to discover new drugs to counter infectious diseases. Conventional wet lab PPI detection techniques have limitations in terms of cost and large-scale application. Hence, computational approaches are developed to predict PPIs. This study aims to develop machine learning models to predict inter-species PPIs with a special interest in HPIs. Specifically, we focus on seeking answers to three

---

[a]This article was previously published in *Journal of Bioinformatics and Computational Biology*. Vol: 16, No. 4 (2018), 1850014 (18 pages).
**Corresponding author.

questions that arise while developing an HPI predictor: (1) How should negative training examples be selected? (2) Does assigning sample weights to individual negative examples based on their similarity to positive examples improve generalization performance? and, (3) What should be the size of negative samples as compared to the positive samples during training and evaluation? We compare two available methods for negative sampling: random versus *DeNovo* sampling and our experiments show that *DeNovo* sampling offers better accuracy. However, our experiments also show that generalization performance can be improved further by using a *soft DeNovo* approach that assigns sample weights to negative examples inversely proportional to their similarity to known positive examples during training. Based on our findings, we have also developed an HPI predictor called HOPITOR (Host-Pathogen Interaction Predictor) that can predict interactions between human and viral proteins. The HOPITOR web server can be accessed at the URL: http://faculty.pieas.edu.pk/fayyaz/software.html#HoPItor.

Keywords: Host–pathogen interactions; interaction predictor; protein–protein interactions; negative sampling.

1. Introduction

Proteins are complex molecules that take part in virtually all life processes in living organisms[1] such as metabolism, signaling, and structural organization.[2] Protein sequences are composed of long chains of 20 amino acids.[1] The sequence of a protein determines its three-dimensional structure, and, consequently, its specific functions.[1,3] Proteins rarely act alone: more than 80% of proteins operate in complexes formed through interactions of proteins.[1,2,4,5] Therefore, to understand functional mechanisms of proteins, it is very important to study the protein–protein interactions (PPIs).[6] A special type of PPIs is host–pathogen interactions (HPIs) that involve interactions between proteins from a pathogen (virus or bacteria) and its host.[7] According to the World Health Organization, each year more than 17 million people are killed by infectious diseases.[8,9] To fight these infectious diseases, it is important to identify HPIs as it is a key step in drug design and biological discovery of disease mechanisms.[6]

Conventional wet lab techniques are expensive and time-consuming, making it almost impossible to assess all possible combinations of protein interactions between a pathogen and its host.[10,11] Therefore, computational approaches are used to predict HPIs.[10,11] For example, if we want to find possible interactions of only 2000 host proteins with 500 pathogen proteins, the possible host-pathogen combinations turn out to be one million. For this reason, there is a shortage of experimentally verified HPI data. Computational studies are vital to increase the available PPI data and eventually add to the pace of drug design research.[11]

Most available computational methods predict HPIs that use sequence and structural similarity-based techniques.[7,11–14] However, due to limited availability of structural information[15] and missing data,[16] sequence-only based methods are better for generalized predictors. In machine learning techniques for PPIs, feature vectors are extracted from protein sequences and then a machine learning model learns from available data to predict unknown interactions.[11–13,17,18] A binary classifier uses a training data set of known interacting and noninteracting protein pairs.[19] The positive samples in training of the classifier are obtained from biochemical experiments, whereas, negative samples are typically generated computationally.[11,13,19,20] Machine learning models for predicting PPIs include Support vector machines (SVMs),[13,17,18] Random Forest (RF),[21] Gradient Boosting Machine (XGBoost),[22] Neural Networks, etc.

In this work, we discuss three questions related to the development of machine learning models for prediction of HPIs, and their performance evaluation as discussed below. We also present a new prediction method called HOPITOR which performs significantly better than previous techniques.

The first question that arises in the development of a machine learning model for HPIs is how to generate negative samples for training a host-pathogen interaction predictor. Ben-Hur *et al.*[20] discuss the available methods to choose negative examples and conclude that negative samples should be selected uniformly at random. Although random sampling may contaminate the negative examples with interacting proteins, this contamination is likely to be minor. More recently, a method called *DeNovo* negative sampling has been proposed by Eid *et al.*[13] to replace random sampling. *DeNovo* sampling is a dissimilarity-based negative sampling criterion that considers sequence similarities of viral proteins interacting with a host protein in generating negative examples.[13] In *DeNovo* sampling, pairs of host and pathogen proteins whose sequences are similar to known positive examples are excluded from the set of negative examples. The experiments by Eid *et al.* show that *DeNovo* sampling is more effective than random sampling in training HPI predictors in terms of generalization performance. However, Eid *et al.*

use a balanced dataset with an equal number of positive and negative proteins selected with *DeNovo* sampling which does not reflect the true use of an HPI prediction method as the number of pairs of host and pathogen proteins can be much larger than possible protein interactions. We use a simple SVM to compare these two methods for sampling negative examples on an imbalanced dataset.

The second question is: whether assigning sample-level weights to negative classification examples based on their similarity to positive examples can help improve generalization performance of HPI predictors or not? Training examples can be assigned sample level weights to reflect domain knowledge or confidence in their labeling or noise contamination. Eid *et al.*[13] and others[17–19] do not assign sample level weights to individual training examples. The idea of using weighted training samples is widely implemented in other fields of science[24–26] Ravikant *et al.*[27] used the idea of weighted samples in an energy-based docking method for PPIs. Inspired by the success of *DeNovo* sampling in selection of negative examples, we hypothesize that assigning sample weights for negative examples are inversely proportional to their similarity to known positive examples that can help in improving classifier generalization. This approach can be thought of as a *soft DeNovo* approach for handling negative examples in training HPI predictors. It generalized the concept of *DeNovo* sampling: negative examples that are similar to known positive examples have a lesser impact on the decision boundary of a classifier in comparison to more dissimilar ones to ensure good generalization. Using such sample level weighting can reduce the impact of false positives during training. We test this hypothesis by comparing the performance of different classifiers with and without sample weights.

The third question considered in this study is: what is the effect of training and test data sizes on prediction performance? Should we use balanced training and evaluation sets with an equal number of positive and negative examples or should we use as much available data as possible? Eid *et al.*[13] and others[17,18] use a balanced set in their evaluation. However, as discussed above, this approach does not simulate real-world use of HPI predictors because the number of possible interactions between proteins is much smaller than all possible pairs of those proteins.[23] Therefore, we propose that the entire data should be used for training an HPI predictor and experimentally show that this approach offers improved prediction performance.

The rest of the paper is organized as follows: Section 2 describes the methods used in the study, Section 3 gives results and discussion about different computational experiments whereas Sec. 4 presents our conclusions.

2. Methods

An overview of the methodology is presented in the flowchart given in Fig. 1. The step-wise development is given below.

2.1. *Datasets and Preprocessing*

HPIs are obtained from supplementary data provided by Eid *et al.*[13] This dataset has originally been compiled using VirusMentha.[28] After removing duplicates, it has 4971 unique interactions between 2237 human proteins and 337 viral proteins. The viral proteins are divided into 10 groups based on their biological families as shown in Table 1. The partitioned data was obtained from Eid *et al.*[13] on request.

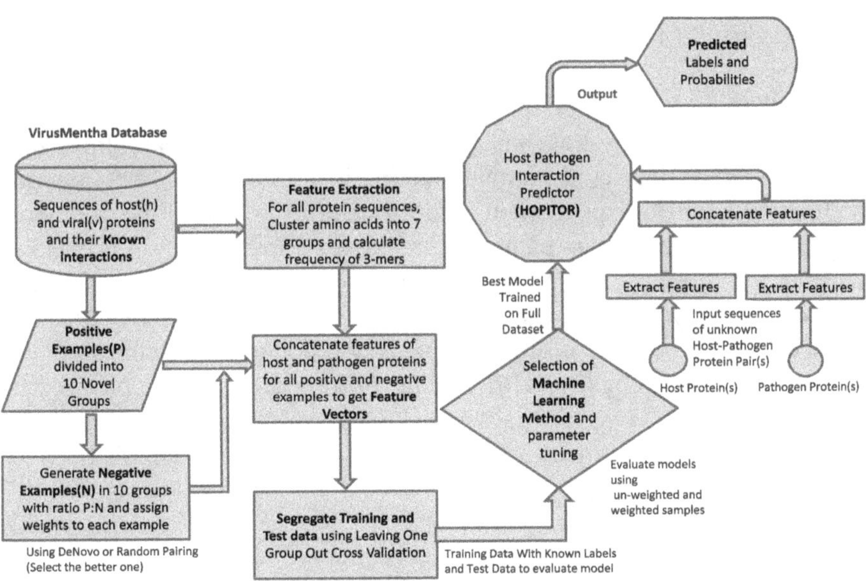

Fig. 1. Flowchart for developing HOPITOR.

Table 1. Partitioned viral families.

Group no.	Family name	No. of viral proteins	Positive examples	*DeNovo* negative examples	Ratio of negative to positive examples
1	Paramyxoviridae	12	762	1797	2.35
2	Filoviridae	4	114	592	5.20
3	Bunyaviridae	3	159	508	3.20
4	Flaviviridae	14	291	25953	89.2
5	Adenoviridae	22	88	3453	39.2
6	Orthomyxoviridae	32	664	5004	7.50
7	Chordopoxviridae	26	194	4158	21.4
8	Papillomaviridae	40	245	6665	27.2
9	Herpesviridae	134	1001	25505	25.5
10	Retroviridae	50	1399	8062	5.76
	Total	**337**	**4917**	**81697**	**16.6**

2.2. Generating Negative Samples

Machine learning-based PPI predictors require both positive and negative datasets for their training. Because the interaction data are available for positive class only, the generation of negative examples is the first step. We evaluate two different techniques for generating negative samples: random versus *DeNovo* sampling.

2.2.1. Random Negative Sampling

Ben-Hur *et al.*[20] argue that, even though random sampling may contaminate the dataset, this contamination is not likely to effect classifier performance much. This method is illustrated in Fig. 2(a). First, a random host protein and a random pathogen protein are chosen; then, it is checked whether the chosen pair is part of the set of known interactions; if not, then the randomly chosen samples are selected as negative examples. The solid connectors represent the positive HPIs and the dotted connectors show the randomly selected negative HPIs. Pathogen protein 1 can be paired with host proteins *a* and *c* as a negative example; however, it cannot be paired negatively with host protein *b* since there exists a positive sample between 1 and *b*.

2.2.2. DeNovo Negative Sampling

Eid *et al.*[13] hypothesize that viral proteins with high sequence similarity can interact with similar host proteins. Based on this hypothesis, the

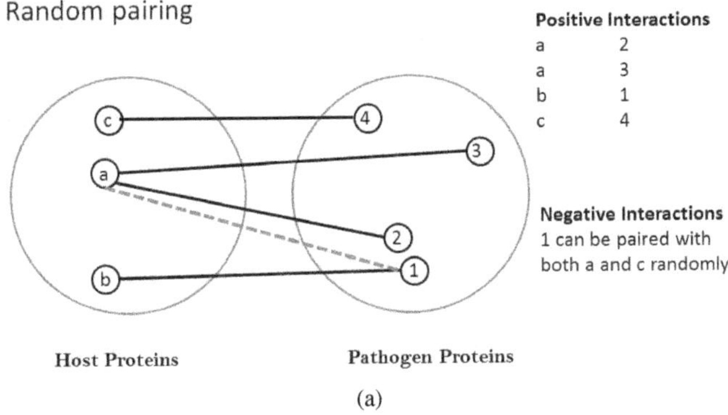

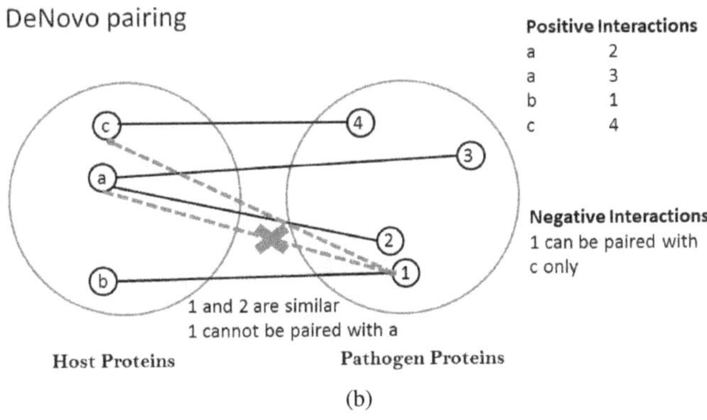

Fig. 2. Illustration of (a) Random sampling and (b) *DeNovo* sampling.

authors argue that random negative sampling will result in a large number of false negatives samples. To mitigate this issue, Eid *et al.*[13] propose dissimilarity-based negative sampling criterion called *DeNovo* negative sampling.

In *DeNovo* sampling, pairwise sequence similarity of viral proteins is first determined. If two viral proteins are more similar than a cut-off value, a host protein that interacts with one of them cannot be paired with the second one to form a negative example. Dissimilarity scores are used to compare the similarities between viral proteins. These scores are obtained by taking the complement of normalized bit scores from the all-versus-all pairwise global alignment of viral proteins. At a dissimilarity threshold T, the negative samples that do not fulfill the criterion are filtered out, and random sampling is done over the rest of the negative examples. We use

$T = 0.7$ in this study. This method is illustrated in Fig. 2(b). The complete details of *DeNovo* pairing technique is given in Eid *et al.*[13] In Fig. 2(b), pathogen protein 1 cannot be paired with host protein b because it interacts with it. Moreover, it also cannot be paired with host protein a as 1 and 2 are similar to each other, i.e. their dissimilarity distance is less than T, and 2 interacts with a. However, viral proteins 1 and 4 have dissimilarity distance $\geq T$, and they have a positive example with human proteins b and c respectively, therefore, 1 can only be paired with c as a negative example. Table 1 shows the number of *DeNovo* negative examples for each viral family in our evaluation. For a fair comparison, an equal number of negative proteins were selected based on random sampling as well.

2.3. *Feature Extraction*

In this work, we have used clustered tripeptide composition features which were also used by Eid *et al.*[13] and originally proposed by Shen *et al.*[17] and others.[12,18] In this approach, the 20 amino acids are first clustered into seven groups based on their physiochemical properties that affect protein interactions such as side chain volume and dipoles.[17] The seven clusters are $\{A, V, G\}$, $\{I, L, F, P\}$, $\{Y, M, T, S\}$, $\{H, N, Q, W\}$, $\{R, K\}$, $\{D, E\}$ and $\{C\}$. After clustering, the frequency of all possible 3-mers is calculated in each protein sequence to get a $7^3 = 343$ dimensional protein-level feature vector. The individual feature vectors in an example (h, v) comprising of a host protein h and pathogen protein v are first normalized to unit norm and concatenated into a single 686-dimensional feature vector.

2.4. *Classification Models*

We evaluate three different regularized machine learning models in our study: SVM, Random Forest (RF) and Gradient Boosting Machine (XGBoost).[22] We use Scikit-learn 0.19[29] in Python 2.7[30] to train and evaluate SVM and RF models and python-based xgboost 0.7 API for training and testing of XGBoost. For notation, we assume that we are given sets of positive and negative examples, P and N, respectively, with a total of $n = |P| + |N|$ examples. An example is represented by its feature vector $\mathbf{x}_i$ and associated label $y_i \in \{+1, -1\}$ for $i = 1, \ldots, n$. Below we discuss classifiers used in this study.

2.4.1. *Support Vector Machines (SVMs)*

We have used SVMs[31] with a radial basis function kernel. SVM is a large-margin classifier with the decision function $f(x) = \mathbf{w}^T x + b$ where $\mathbf{w}$ and b are weight and bias terms of the classifier. The optimization problem of an SVM tries to minimize the number of margin violations and misclassifications over training data while maximizing its regularization or margin:

$$\min_{\mathbf{w},b,\xi} \frac{1}{2}\|\mathbf{w}\|^2 + \sum_{i=1}^{n} c_i \xi_i, \tag{1}$$

$$\text{such that for all } i = 1,\ldots,n : y_i(\mathbf{w}^T \mathbf{x}_i + \mathbf{b}) \geq 1 - \xi_i, \quad \xi_i \geq 0. \tag{2}$$

Here, ξ_i is the extent of margin error for the ith example. The hyper-parameter c_i is the margin violation penalty of the ith example and it determines the relative significance of its margin violation and maximization of the margin. Typically, the margin violation penalties of all examples are set equal to C to reduce the number of hyper-parameters. However, these margin violation penalties can be used to assign sample weights to individual examples as well. The above equation can be expressed as the dual form:

$$\max_{\alpha} \sum_{i=1}^{n} \alpha_i - \frac{1}{2} \sum_{i,j=1}^{n} y_i y_j \alpha_i \alpha_j \mathbf{x}_i^T \mathbf{x}_j, \tag{3}$$

$$\text{such that} : \sum_{i=1}^{n} y_i \alpha_i = 0, \quad \text{and,} \quad 0 \leq \alpha_i \leq c_i. \tag{4}$$

In (3), the dot product term $\mathbf{x}_i^T \mathbf{x}_j$ can be replaced by a generalized dot product or kernel function, $K(\mathbf{x}_i, \mathbf{x}_j)$ to make the classifier nonlinear. In this study, radial basis function is used which is given as: $K(\mathbf{x}_i, \mathbf{x}_j) = \exp(-\gamma\|\mathbf{x}_j - \mathbf{x}_j\|^2)$. We optimize the SVM kernel parameters using grid search and select $C = 10, \gamma = 0.1$.

2.4.2. *Random Forest (RF)*

Random Forest (RF) is an ensemble learning technique that works by creating several decision trees on arbitrary subsample sets of input features during training. In testing, it produces a probability value for a test example to belong to the positive class by taking the mode of classes of individual trees.[21] We optimize the hyper-parameters of the RF ensemble using grid search with respect to maximum number, depth and splitting features of decision trees.

2.4.3. *Gradient Boosting Machine (XGBoost)*

XGBoost is an ensemble learning technique that operates by combining weak decision tree learners in an iterative manner through boosting. This technique uses gradient boosting to learn a decision function that minimizes the average value of classification loss on training data based on a combination of decision trees trained in an incremental manner over residual errors of the previous stage.[32]

We use the grid search to optimize XGBoost parameters with respect to maximum depth, objective function, learning rate, booster subsample and number of boosting iterations. The optimal parameters are: learning rate $= 0.1$, max depth $= 10$, no. of estimators $= 100$ and subsample $= 1$.

2.5. *Soft DeNovo: Weighting of Negative Examples in Training*

In training machine learning models for classification, training examples can be assigned a sample weight based on confidence about the correctness of their labels, the degree of noise in the sample or class sizes. Sample weighting can be used to include domain-specific knowledge about examples in training a machine learning model. For example, in a support vector machine, each example can be assigned its own margin violation penalty c_i instead of having a global C hyper-parameter as discussed earlier. One advantage that this flexibility allows is to handle class imbalance by assigning a larger margin violation penalty to the minority class examples in comparison to examples from the majority class. Similar sample weighting schemes are also available for both random forest and XGBoost classifiers as well.[21,32]

In this work, we propose to use sample weights to reflect our confidence in the correctness of labels of negative training examples based on their sequence similarity to positively labeled training examples. To test our hypothesis that sample weighting based on this approach will improve classification performance, we compare various classifiers with and without sample weighting as discussed below.

2.5.1. *Unweighted Training Samples*

In this approach,[31] all examples are weighted equally during training. However, to handle class imbalance, all examples in a class are assigned the same weight which is inversely proportional its prior class probability.

2.5.2. *Soft DeNovo*

In this approach,[24] a separate value of sample weight is assigned to each training example. As positive training examples are experimentally validated, therefore, all positive examples are assigned a sample weight of 1.0. On the other hand, negative training examples are computationally generated, therefore, negative examples are assigned weights between 0 and 1 depending upon their sequence similar to known positive examples. Specifically, we set the sample weight of a negative example comprising of a host protein h and viral protein v to the dissimilarity score of v with its most similar viral protein that interacts with the host protein h. This idea is illustrated in Fig. 3. Mathematically, the sample weight of a negative example consisting of a host protein h and viral protein v can be written as: $\min_{v' \in \{v'' | (h,v'') \in P\}} D_{vv'}$ where $D_{vv'}$ is the normalized dissimilarity score between viral proteins v and v'. This method of assigning sample weights can be thought of as a *soft DeNovo* approach for handling negative examples. It generalizes the *DeNovo* sampling concept proposed by Eid *et al.* negative examples that have high sequence similarity with positive examples should have a smaller effect on determining the

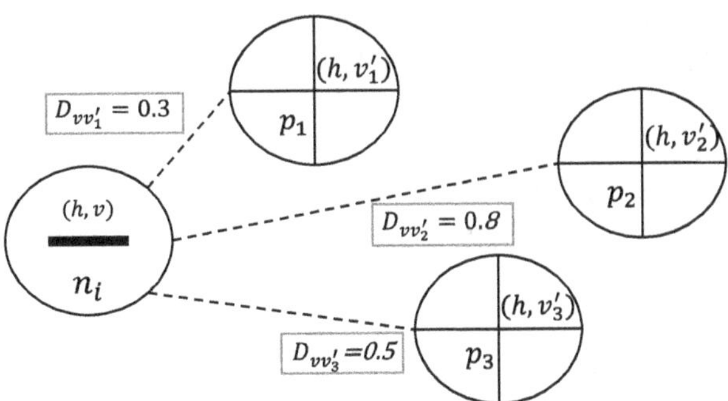

Fig. 3. An example of selecting the weight, c_i of a negative sample. Consider the pair $(\boldsymbol{h}, \boldsymbol{v})$ between host protein $\boldsymbol{h}$ and viral protein $\boldsymbol{v}$ such that the host protein $\boldsymbol{h}$ interacts with v'_1, v'_2, and v'_3. Similarity of $\boldsymbol{v}$ is maximum with v'_1. Therefore, dissimilarity distance $\boldsymbol{D_{vv'_1}} = \boldsymbol{0.3}$ is set as the sample weight c_i.

classification boundary in comparison to more dissimilar ones to ensure good generalization by reducing the impact of potential false positives. This allows the classifier to construct a decision boundary through an area of low data density. Assigning weights to negative samples in this manner can also reduce the impact of the method used for selecting negative examples on generalization performance of an HPI predictor. To handle class imbalance, the sample weight of an example is then multiplied by its class-level weight which is inversely proportional to its prior class probability.

2.6. *Performance Evaluation*

Machine learning models in this study have been evaluated using leave-one-group-out cross-validation as implemented by Eid *et al.*[13] In this evaluation, a machine learning model is trained on examples from all but one of the 10 groups listed in Table 1 and tested on the held-out group or family. This gives a more realistic assessment of generalization performance in scenarios where the predictor will be used for identifying protein interactions of a novel viral family.

The cross-validation performance of different machine learning models is evaluated using area under the Receiver Operating Characteristic curve (AUC-ROC) and precision-recall curves (AUC-PR).[11,33] For this purpose, True positive rate (TPR) (or recall), False Positive Rate (FPR) and precision are defined as follows: $\text{TPR} = \frac{\text{TP}}{\text{TP+FN}}$, $\text{FPR} = \frac{\text{FP}}{\text{TP+FN}}$, $\text{Precision} = \frac{\text{TP}}{\text{TP+FP}}$.

AUC-ROC is the area under the plot between TPR and FPR at various thresholds. The ROC curve tells us how good a predictor is at detecting true positives at a given rate of false positives. The ROC curve is not sensitive to class imbalance. To report our results on imbalanced datasets, we have used the area under the precision-recall curve (AUC-PR). The weighted averages of AUC-ROC and AUC-PR for all groups with respect to the number of examples in them are reported for comparison.

2.7. *Biological Validation*

To test the generalization performance of our prediction model, we test it on three different viral species (Human respiratory syncytial virus (HRSV), Measles virus, and Rabies virus) from three different families. Specifically, we use our trained model to identify the interactions of all proteins in a viral proteome with human STAT1 and STAT2 proteins. It is important to note that these examples are not part of our training data and one of the viral families is completely novel for the classifier. We rank all putative interactions of the viral proteins with their human host proteins and compare them to the literature as discussed in the results section.

3. Results and Discussion

3.1. *Negative Sampling*

In order to select the optimal method for generating negative examples, we set-up two SVM models to test Random and *DeNovo* negative sampling. The first model is trained with negative examples generated by using Random negative sampling whereas the second model is trained with negative examples generated by using *DeNovo* negative sampling at $T = 0.7$. Eid *et al.*[13] use the same number of positive and negative examples. However, to simulate the real word scenario where the number of positive examples is expected to be much smaller than negative examples, we select the entire set of *DeNovo* negative examples instead of using the balanced set. Both classifiers are tested on positive and *DeNovo* negative examples using leave one group out cross-validation. The weighted average AUC-ROC for the model trained with random negative sampling is 0.47 whereas, the AUC-ROC for the model trained with *DeNovo* negative sampling is 0.68. Moreover, the average AUC-PR for random sampling is 0.05, while it is 0.39 for *DeNovo* sampling.

The detailed group-wise results are shown in Table 2. These significantly improved results show that *DeNovo* sampling is indeed better than Random sampling. Therefore, we choose *DeNovo* negative sampling for building our HPI predictor and comparing classifiers as discussed in the next section. This result is in agreement with the findings of Eid *et al.* However, unlike the work by Eit *et al.*, we have used an imbalanced dataset with precision-recall scores. As a consequence, our result further generalizes the conclusion that *DeNovo* sampling should be used in training HPI predictors.

3.2. *Classifier Comparison*

We compare the performance of three different classifiers (SVM, RF, and XGBoost). The cross-validation results of these three classifiers are

Table 2. Comparison of random versus *DeNovo* sampling for SVM classifier.

Group	Random sampling		*DeNovo* sampling	
	AUC-ROC	**AUC-PR**	**AUC-ROC**	**AUC-PR**
1	0.43	0.244	0.96	0.95
2	0.272	0.101	0.984	0.973
3	0.38	0.203	0.952	0.943
4	0.53	0.015	0.277	0.011
5	0.487	0.027	0.989	0.917
6	0.405	0.09	0.877	0.736
7	0.412	0.03	0.962	0.778
8	0.437	0.023	0.932	0.661
9	0.478	0.034	0.816	0.358
10	0.353	0.106	0.773	0.517
Score	**0.47**	**0.05**	**0.68**	**0.39**

Table 3. Comparison of SVM, RF and XGBoost for *DeNovo* sampling.

Group	SVM		RF		XGBoost	
	ROC	**PR**	**ROC**	**PR**	**ROC**	**PR**
1	0.96	0.95	0.998	0.996	0.999	0.999
2	0.984	0.973	0.998	0.99	1.000	1.000
3	0.952	0.943	0.999	0.995	1.000	1.000
4	0.277	0.011	0.269	0.008	0.305	0.01
5	0.989	0.917	0.999	0.977	0.999	0.99
6	0.877	0.736	0.957	0.825	0.999	0.998
7	0.962	0.778	0.998	0.979	0.999	0.994
8	0.932	0.661	0.962	0.653	0.994	0.805
9	0.816	0.358	0.894	0.345	0.906	0.52
10	0.773	0.517	0.951	0.844	0.97	0.911
Score	**0.68**	**0.39**	**0.73**	**0.44**	**0.76**	**0.53**

tabulated in Table 3. It can be seen that XGBoost outperforms both SVM and RF both in terms of AUC-ROC and AUC-PR. As a consequence, we use XGBoost for our final model.

3.3. *Effect of Sample Weighting*

In order to analyze the effect of sample weighting, we compare the three classification techniques (SVM, RF, and XGBoost) with and without

Table 4. Comparison of SVM, RF, and XGBoost with weighted training samples. The last row is added from Table 3 for easier comparison. The best performance scores are highlighted in bold.

Classifier	SVM		RF		XGBoost	
Viral family group	**ROC**	**PR**	**ROC**	**PR**	**ROC**	**PR**
1	0.994	0.992	0.996	0.989	0.999	0.999
2	0.999	0.998	0.995	0.979	1.000	1.000
3	0.970	0.964	0.998	0.993	1.000	1.000
4	0.269	0.011	0.274	0.008	0.289	0.008
5	0.995	0.957	0.997	0.936	0.999	0.991
6	0.964	0.908	0.974	0.877	0.999	0.994
7	0.986	0.932	0.994	0.961	0.998	0.988
8	0.972	0.722	0.98	0.743	0.994	0.870
9	0.910	0.570	0.927	0.633	0.925	0.620
10	0.896	0.667	0.962	0.863	0.957	0.863
With sample weighting	**0.73**	**0.5**	**0.75**	**0.54**	**0.76**	**0.56**
Without sample weighting	0.68	0.39	0.73	0.44	**0.76**	0.53

sample weighting. The results are shown in Table 4 and are discussed below. It is interesting to note that sample weighting improves prediction performance for all classification techniques especially in terms of the Area under the PR curve. This confirms our hypothesis that the proposed soft *DeNovo* technique for selecting negative examples improves prediction performance by reducing the impact of potential false positives during training.

For SVM, the AUC-PR score for un-weighted training samples is 0.39. It improves to 0.5 when sample weights are used in training. Similarly, the AUC-ROC score improves from 0.68 (without sample weights) to 0.73 with sample level weighting. For RF, the AUC-ROC score for weighted training samples is 0.75 as compared to 0.73 for un-weighted samples. Similarly, the AUC-PR improves from 0.44 to 0.54 as a consequence of sample weighting. For XGBoost, the weighted AUC-PR improved from 0.53 to 0.56 when the classifier was trained with weighted samples. However, the weighted AUC-ROC remained the same for both un-weighted and weighted training samples. It is also important to note that accuracy of group 4 is consistently low for all classifiers. This is because the viral proteins in this group belonging to Flaviviridae, are very dissimilar to the rest of the protein families.

3.4. *Effect of Training and Test Data Sizes*

We evaluate the performance of our best performing classifier (XGBoost) with *DeNovo* sampling of negative examples with sample level weights using a balanced ($|P| = |N| = 4971$) and the full data set ($|P| = 4971$, $|N| = 81{,}697$). The objective of this experiment is to see the effect of training and test data sizes on classification performance. The negative examples in the balanced set are a random subset of examples from the set of 81,697 *DeNovo* negative examples stratified with respect to their groups. The results of leave one group out cross validation on all training and test combinations of balanced versus full data sets are given in Table 5. It is interesting to note that restricting testing to the balanced test set gives much higher performance scores in comparison to the more realistic full dataset testing. This clearly shows that restricting to a balanced test set for performance evaluation can lead to inflated accuracy values. Furthermore, using the full training set with sample weighting can improve prediction performance over the full test set in comparison to training the classifier on a balanced training set. Therefore, we conclude that HPI predictors should be trained and evaluated using the full set of negative examples rather than restricting to a balanced set. Not only does this give a more realistic assessment of the real world use of HPI predictors, but it can also lead to better generalization performance.

3.5. *Web Server Implementation*

We train our final HPI predictor using XGBoost on the full dataset using *DeNovo* negative sampling and sample level weighting. We have developed a web server of our HPI predictor for the biologists to check whether a human protein interacts with a viral protein or not. It is called Host-Pathogen Interaction Predictor (HOPITOR). It can be accessed

Table 5. Comparison of AUC-PR and AUC-ROC (in parenthesis) scores for XGBoost trained and tested with balanced and full datasets using leave one group out cross-validation. The best performance scores are highlighted in bold.

	Balanced training set	**Full training set**
Balanced test set	0.95 (0.94)	0.95 (0.94)
Full test set	0.50 (0.75)	0.56 (0.76)

through http://faculty.pieas.edu.pk/fayyaz/software.html# HoPItor for free. Due to computational requirements, the server is limited to a single protein pair for testing at one time by a single user. However, we also provide the source code of the method for large-scale use.

3.6. *In Silico Predictions and Biological Validation*

We use HOPITOR for *in silico* predictions for biologically validated host-pathogen interactions that are not a part of our training set to ensure its generalization performance. The sequences of proteins discussed below are taken from the UniProt[34] repository. Specifically, we use three different viral proteomes to identify their interactions with human STAT1 and STAT2 proteins which are part of the human immune response to pathogens.

3.6.1. *Human STAT2 and HRSV Proteins*

Human respiratory syncytial virus (HRSV, family: *Paramyxoviridae*) is the main cause of lower respiratory tract infections.[35,36] HRSV has 11 proteins out of which nonstructural proteins (NS1 and NS2) specifically degrade human STAT2 protein.[37,38] We tested all proteins in the HRSV proteome for interaction with human STAT2 protein on HOPITOR. The predicted probability of interaction between NS1 and human STAT2 is 99% and it ranks at the top among all 11 viral proteins. HOPITOR also predicts 90% probability of NS2 and human STAT2 proteins to interact with each other.

3.6.2. *Human STAT1 and Measles Proteins*

Measles virus (family: *Paramyxoviridae*) phosphoprotein P blocks the phosphorylation of human STAT1 protein by interacting with it.[39,40] We tested the interactions between all proteins in the Measles virus proteome and human STAT1. The predicted probability of interaction between STAT1 and measles virus protein P is 96.1%. This interaction is ranked second among seven proteins of measles virus. The highest ranked protein is nucleoprotein with probability 96.5%.

3.6.3. *Human STAT1 and Rabies Proteins*

Rabies virus (family: *Rhabdoviridae*) phosphoprotein interacts with human STAT1 protein to inhibit interferon signal transduction pathways.[41]

We tested human STAT1 protein for interaction with all proteins in the rabies virus proteome. The predicted probability of interaction between STAT1 with rabies phosphoprotein is 91.4%. This prediction is ranked second among five proteins of rabies virus. It is important to note that *Rhabdoviridae* family was not part of our training set.

These *in silico* prediction experiments show that HOPITOR can be used as a reliable tool to predict human and viral protein interactions.

4. Conclusions

In this paper, we tried to answer three questions that arise while developing a host-pathogen protein–protein interaction predictor. Below are the conclusions from our findings and contributions:

(1) *DeNovo* sampling is better than random sampling for generating negative examples. We suggest that HPI predictors be trained using *DeNovo* negative examples.
(2) We propose a *soft DeNovo* approach for handling negative examples in training HPI predictors that offers better generalization performance by assigning lower weights to negative training examples that are similar to positive ones.
(3) HPI predictors should be trained using the complete set of negative examples instead of restricting to a balanced dataset.

We have also developed a new HPI predictor called HOPITOR using the XGBoost classifier based on these findings. Our computational results show that HOPITOR can be used for practical applications.

Acknowledgments

The authors thank Fatma-Elzahraa Eid, Virginia Virginia Polytechnic Institute and State University, USA for providing the relevant data for this study and continuous support throughout this study. We are also grateful to Dr. Brian Geiss at Colorado State University, USA in assisting us with the biological validation of our approach.

Abdul Hannan Basit and Amina Asif are supported by PIEAS MS Fellowship and IT and Telecom Endowment Fund, respectively. Wajid A. Abbasi and Sadaf Khan are supported by grant under indigenous 5000 Ph.D. fellowship scheme from the Higher Education Commission

(HEC) of Pakistan (PIN: 213-58990-2PS2-046, PIN: 315-12753-2EG3-197). This work is also supported by National Research Program for Universities (NRPU), Higher Education Commission, Pakistan (Project ID: 6085).

References

1. Kessel A, Ben-Tal N, *Introduction to Proteins: Structure, Function, and Motion*, CRC Press, 2010.
2. Braun P, Gingras A-C, History of protein-protein interactions: From egg-white to complex networks, *Proteomics* **12**:1478–1498, 2012.
3. Rigden DJ, *From Protein Structure to Function with Bioinformatics*, Springer Science & Business Media, 2008.
4. Berggård T, Linse S, James P, Methods for the detection and analysis of protein-protein interactions, *Proteomics* **7**:2833–2842, 2007.
5. Waugh DF, Protein-Protein Interactions, in Anson ML, Bailey K, Edsall JT (eds.), *Advances in Protein Chemistry*, Academic Press, pp. 325–437, 1954.
6. Dömling A, *Protein-Protein Interactions in Drug Discovery*, John Wiley & Sons, 2013.
7. Dyer MD, Murali TM, Sobral BW, Computational prediction of host-pathogen protein–protein interactions, *Bioinformatics* **23**:i159–i166, 2007.
8. WHO Press release. *WHO* Available at http://www.who.int/whr/1996/media_centre/press_release/en/. Accessed: 16th October 2016.
9. WHO Press release. *WHO* Available at http://www.who.int/whr/1996/media_centre/press_release/en/index1.html. Accessed: 16th October 2016.
10. Pitre S *et al.*, Computational methods for predicting protein–protein interactions, in Werther M, Seitz H (eds.), *Protein–Protein Interaction*, Springer, Berlin Heidelberg, pp. 247–267, 2008.
11. Nourani E, Khunjush F, Durmus S, Computational approaches for prediction of pathogen-host protein-protein interactions, *Front. Microbiol.* **6**:94, 2015.
12. Dyer MD, Murali TM, Sobral BW, Supervised learning and prediction of physical interactions between human and HIV proteins, *Infect Genet Evol* **11**:917–923, 2011.
13. Eid F-E, ElHefnawi M, Heath LS, DeNovo: Virus-host sequence-based protein–protein interaction prediction, *Bioinformatics* **32**:1144–1150, 2016.
14. Davis FP, Barkan DT, Eswar N, McKerrow JH, Sali A, Host–pathogen protein interactions predicted by comparative modeling, *Protein Sci* **16**:2585–2596, 2007.
15. Berman H, Henrick K, Nakamura H, Markley JL, The worldwide protein data bank (wwPDB): Ensuring a single, uniform archive of PDB data, *Nucleic Acids Res* **35**:D301–D303, 2007.

16. Kshirsagar M, Carbonell J, Klein-Seetharaman J, Techniques to cope with missing data in host–pathogen protein interaction prediction, *Bioinformatics* **28**:i466–i472, 2012.

17. Shen J *et al.*, Predicting protein–protein interactions based only on sequences information, *Proc Natl Acad Sci USA* **104**:4337–4341, 2007.

18. Cui G, Fang C, Han K, Prediction of protein-protein interactions between viruses and human by an SVM model, *BMC bioinformatics* **13**:1, 2012.

19. Abbasi WA, Minhas FUAA, Issues in performance evaluation for host-pathogen protein interaction prediction, *J Bioinform Comput Biol* **14**:1650011, 2016, doi: 10.1142/S0219720016500116.

20. Ben-Hur A, Noble WS, Choosing negative examples for the prediction of protein-protein interactions, *BMC Bioinformatics* **7**:1–6, 2006.

21. Breiman L, Random forests, *Mach Learn* **45**:5–32, 2001.

22. Chen T, Guestrin C, *XGBoost: A Scalable Tree Boosting System*, ACM Press, 2016, pp. 785–794, doi: 10.1145/2939672.2939785.

23. Park Y, Marcotte EM, Revisiting the negative example sampling problem for predicting protein–protein interactions, *Bioinformatics* **27**:3024–3028, 2011.

24. Wu Y, Liu Y, Adaptively weighted large margin classifiers, *J Comput Graph Stat* **22**:416–432, 2013.

25. Chen H, Wang J, Yan X, A fuzzy support vector machine with weighted margin for flight delay early warning, *Fifth Int Conf Fuzzy Systems and Knowledge Discovery, 2008 (FSKD '08)*, pp. 331–335, 2008.

26. Yang X, Song Q, Wang Y, A weighted support vector machine for data classification, *Int J Pattern Recognit Artif Intell* **21**:961–976, 2007.

27. Ravikant DVS, Elber R, Energy design for protein-protein interactions, *J Chem Phys* **135**(6):065102, 2011, doi: 10.1063/1.3615722.

28. Calderone A, Licata L, Cesareni G, VirusMentha: A new resource for virus-host protein interactions, *Nucleic Acids Res* **43**:D588–D592, 2015.

29. Pedregosa F *et al.*, Scikit-learn: Machine learning in python, *J Mach Learn Res* **12**:2825–2830, 2011.

30. Welcome to Python.org. *Python.org* Available at: https://www.python.org/. Accessed: 16th October 2016.

31. Ben-Hur A, Weston J, A user's guide to support vector machines, *Methods Mol Biol* **609**:223–239, 2010.

32. Friedman JH, Greedy function approximation: A gradient boosting machine, *Ann Statist* **29**:1189–1232, 2010.

33. Davis J, Goadrich M, The relationship between precision-recall and ROC curves, *Proc 23rd Int Conf Machine Learning*, pp. 233–240, 2006.

34. UniProt. Available at: http://www.uniprot.org/. Accessed: 12th May 2016.

35. Sigurs N, Epidemiologic and clinical evidence of a respiratory syncytial virus-reactive airway disease link, *Am J Respir Crit Care Med* **163**:S2–S6, 2001.

36. Tripp RA, Oshansky C, Alvarez R, Cytokines and respiratory syncytial virus infection, *Proc Am Thorac Soc* **2**:147–149, 2005.
37. Elliott J *et al.*, Respiratory syncytial virus NS1 protein degrades STAT2 by using the Elongin-Cullin E3 ligase, *J Virol* **81**:3428–3436, 2007.
38. Lo MS, Brazas RM, Holtzman MJ, Respiratory syncytial virus nonstructural proteins NS1 and NS2 mediate inhibition of stat2 expression and alpha/beta interferon responsiveness, *J Virol* **79**:9315–9319, 2005.
39. Devaux P, Priniski L, Cattaneo R, The measles virus phosphoprotein interacts with the linker domain of STAT1, *Virology* **444**:250–256, 2013.
40. Caignard G *et al.*, Measles virus V protein blocks Jak1-mediated phosphorylation of STAT1 to escape IFN-α/β signaling, *Virology* **368**:351–362, 2007.
41. Vidy A, Chelbi-Alix M, Blondel D, Rabies virus P protein interacts with STAT1 and inhibits interferon signal transduction pathways, *J Virol* **79**:14411–14420, 2005.

Abdul Hannan Basit is currently working for Larkana Institute of Nuclear Medicine & Radiotherapy (LINAR), Larkana, Pakistan as a biomedical engineer. Basit received his MS degree in Systems Engineering from Pakistan Institute of Engineering and Applied Sciences (PIEAS), Islamabad, Pakistan under PIEAS MS Fellowship program. His MS thesis research focuses on "Machine learning in Biomedical Informatics".

Wajid Arshad Abbasi is currently working as a visiting research scholar under the International Research Support Initiative Program (IRSIP) of the Higher Education Commission (HEC) of Pakistan, to carry out part of his PhD research at the Department of Computer Science at Colorado State University (CSU), USA. He is pursuing his PhD in the Department of Computer and Information Sciences, Pakistan Institute of Engineering and Applied Sciences (PIEAS), Islamabad, Pakistan. He has also been awarded the indigenous PhD fellowship by the HEC. His primary area of research is "Machine Learning in Bioinformatics". His webpage is available at http://faculty.pieas.edu.pk/fayyaz/wajid/index.html.

Amina Asif is a PhD student at the Department of Computer and Information Sciences, Pakistan Institute of Engineering and Applied Sciences (PIEAS), Islamabad, Pakistan. Her primary area of research is Applied Machine Learning.

Sadaf Gull is a PhD scholar in the Department of Computer and Information Sciences, Pakistan Institute of Engineering and Applied Sciences (PIEAS), Islamabad, Pakistan. She is doing her PhD under indigenous PhD fellowship by Higher Education Commission (HEC). Her area of research is "Machine Learning in Biomedical Informatics".

 **Fayyaz Ul Amir Afsar Minhas** is a principal scientist with the Department of Computer and Information Sciences, Pakistan Institute of Engineering and Applied Sciences (PIEAS), Islamabad, Pakistan. Dr. Minhas received his PhD degree in Bioinformatics from Colorado State University, USA on a Fulbright Scholarship. He has also been awarded the National Youth Award by the Government of Pakistan for his contributions to science and technology. His lab focuses on applications of machine learning in Bioinformatics and the analysis of biomedical data. His webpage is available at http://faculty.pieas.edu.pk/fayyaz/.

CHAPTER 7

An Improved Method for Predicting Interactions Between Virus and Human Proteins

Byungmin Kim[*,‡], Saud Alguwaizani[*,§], Xiang Zhou[*,¶],
De-Shuang Huang[†,ǁ], Byunkyu Park[*,**] and Kyungsook Han[*,††]

*Department of Computer Science and Engineering,
Inha University, Incheon 22212, South Korea*

†*Machine Learning and Systems Biology Lab,
College of Electronics and Information Engineering,
Tongji University, Shanghai 201804, P. R. China*
‡*bk86@inha.edu*
§*salguwaizani@inha.edu*
¶*jusang486@gmail.com*
ǁ*dshuang@tongji.edu*
****bpark@inha.ac.kr*
††*khan@inha.ac.kr*

The interaction of virus proteins with host proteins plays a key role in viral infection and consequent pathogenesis. Many computational methods have been proposed to predict protein–protein interactions (PPIs), but most of the computational methods are intended for PPIs within a species rather than PPIs across different species such as virus–host PPIs. We developed a method that represents key features of virus and human proteins of variable length into a feature vector of fixed length. The key features include the relative frequency of amino acid triplets (RFAT), the frequency difference of amino acid triplets (FDAT) between virus and

[a]This article was previously published in *Journal of Bioinformatics and Computational Biology*. Vol: 15, No. 1 (2017), 1650024 (17 pages).
††Corresponding author.

host proteins, and amino acid composition (AC). We constructed several support vector machine (SVM) models to evaluate our method and to compare our method with others on PPIs between human and two types of viruses: human papillomaviruses (HPV) and hepatitis C virus (HCV). Comparison of our method to others with same datasets of HPV–human PPIs and HCV–human PPIs showed that the performance of our method is significantly higher than others in all performance measures. Using the SVM model with gene ontology (GO) annotations of proteins, we predicted new HPV–human PPIs. We believe our approach will be useful in predicting heterogeneous PPIs.

Keywords: Protein–protein interactions; human papillomaviruses; hepatitis C virus.

1. Introduction

Every year millions of people suffer from viral infections caused by a variety of viruses, and the improvement of prevention and treatment of viral infections and associated diseases remains one of the main public health challenges.[1] Ebola virus, Middle East respiratory syndrome coronavirus (MERS-CoV), HIV-1, influenza virus, human papillomavirus (HPV), herpes virus, and hepatitis A, B, C, D, and E viruses are well-known examples of viruses that cause human diseases. During viral infection viruses interact with a large number of cellular proteins, and thus deciphering interactions between virus and human proteins offers an insight into the mechanism of viral infections and can help to design antiviral drugs.[2–5] Recently, proteome-wide studies of viral interactions with human proteins were performed, but comprehensive analysis of the interactions between virus proteins and human proteins involved in viral infections has not been carried out.

So far many computational methods have been proposed to predict protein–protein interactions (PPIs), but most of the methods are intended for PPIs within a same species rather than for PPIs across different species.[21] Methods designed for predicting intra-species PPIs are not appropriate for predicting inter-species PPIs since such methods do not distinguish interactions between proteins of the same species from those of different species. Zhou *et al.*[6,7] showed that a stringent homology-based method is better than a conventional homology-based method in predicting host–pathogen PPIs.

Due to a recent increase in data of virus–host PPIs, a few computational methods have been developed to predict virus–host PPIs using machine learning methods. Cui *et al.*[8] developed a support vector machine (SVM) model using the relative frequency of three consecutive amino acids in a protein sequence. Cui's SVM model was tested to predict the interactions between two types of viruses (hepatitis C virus (HCV) and HPV) and human proteins, and achieved an average accuracy above 80%, which is higher than those of previous methods. Barman *et al.*[9] used three machine learning methods (SVM, Naïve Bayes, and Random Forest) to predict virus–host PPIs using several features such as domain–domain association in interacting protein pairs and composition of methionine, serine, and valine in viral proteins. In a 5-fold cross-validation with the dataset from VirusMINT,[10] their SVM showed higher sensitivity and F1 score than NB and RF, but its accuracy is lower than that of Cui's SVM model. Emamjomeh *et al.*[11] developed an ensemble learning method to predict PPIs between human and HCV proteins using six features of amino acid composition (AC), pseudo AC, evolutionary information feature, network centrality measures, tissue information, and posttranslational modification information. They used multilayer perceptron (MLP) as a meta-learner to combine the predictions of RF, NB, and SVM. In comparison with Cui's SVM model with the same dataset of HCV–human PPIs, Emamjomeh's method showed an accuracy of 83%, which is marginally higher than the average accuracy of 81.6% of Cui's model.

In this study, we constructed several SVM models to predict virus–human PPIs from sequence data using different combinations of three features: the relative frequency of amino acid triplets (RFAT) in virus and host proteins, frequency difference of amino acid triplets (FDAT) between virus and host proteins, and AC in a pair of virus and host proteins. AC is commonly used in predicting PPIs. However, RFAT and FDAT between virus and host proteins were developed in our study to predict PPIs between virus and host. In both cross-validations and independent testing with an extensive datasets of HPV–human PPIs and HCV–human PPIs, the SVM models showed significant improvement in all performance measures (sensitivity, specificity, accuracy, positive predictive value, negative predictive value, and Matthews correlation coefficient) by combining all three features. The rest of this paper discusses the features, representation of proteins, and comparison with other computational methods.

2. Methods

2.1. *Features and Representation of Proteins*

Since proteins involved in PPIs are of different lengths, finding an appropriate way of representing them in a feature vector of the same length is crucial to the success of computational methods for predicting PPIs.

Several methods have been developed to represent protein sequences for predicting PPIs. For example, Shen *et al.*[12] clustered 20 amino acids into seven classes, {AVG}, {ILFP}, {YMTS}, {HNQW}, {RK}, {DE}, and {C} based on the diploes and volumes of the side chains of amino acids. The frequency of three consecutive amino acids (what we call *amino acid triplet*) in a protein sequence was represented using the classification. Since there are $7 \times 7 \times 7 = 343$ possible amino acid triplets, a feature vector of 686 elements (343 for one protein and 343 for another protein) was generated for each protein pair. Each element of a feature vector represents the relative frequency d_i of an amino acid triplet i in the protein sequence, which is defined by Eq. (1). In the equation, f_i is the frequency of the ith triplet, $\max\{f_1, f_2, \ldots, f_{343}\}$ is the maximum frequency, and $\min\{f_1, f_2, \ldots, f_{343}\}$ is the minimum frequency in the protein sequence.

$$d_i = \frac{f_i - \min\{f_1, f_2, \ldots, f_{343}\}}{\max\{f_1, f_2, \ldots, f_{343}\} - \min\{f_1, f_2, \ldots, f_{343}\}}. \tag{1}$$

Cui *et al.*[8] improved Shen's representation to predict virus–host PPIs. Cui's method clustered 20 amino acids into six categories, {IVLM}, {FYW}, {HKR}, {DE}, {QNTP}, and {ACGS} based on their biochemical properties, and redefined the relative frequency of amino acid triplet by Eq. (2).

$$d_i = \left\{ e^{\frac{f_i - \min\{f_1, f_2, \ldots, f_{216}\}}{\max\{f_1, f_2, \ldots, f_{216}\} - \min\{f_1, f_2, \ldots, f_{216}\}}} \right\} - 1. \tag{2}$$

For each pair of virus and human proteins, Cui *et al.* generated a feature vector of 217 elements ($6 \times 6 \times 6 = 216$ for a human protein and 1 for the type of a virus protein). While the relative frequency in Shen's representation has a value in the range [0, 1], Cui's relative frequency has a value in a wider range [0, 1.718], which makes it easier to discriminate proteins. In addition, the 686-dimensional vector generated by Shen's method was reduced to about 1/3 (217-dimensional vector) in Cui's method, which in general contributes to reduce the training time and space of a prediction model.

However, our experience shows that both Shen's and Cui's representations generate a feature vector with many zero-valued elements. This is because the minimum frequency of amino acid triplets is zero in most proteins, and thus $f_i - \min\{f_1, f_2, \ldots, f_M\}$ becomes zero for every amino acid triplet that is not present in a protein. Zero-valued elements remain zeros after scaling. Thus, feature vectors with too many zero elements do not capture much information, which often make the performance of a classifier low.

The problem with too many zero-valued elements in feature vectors was also noted by Sharma *et al.*[13] Since feature vectors with many zero elements lower the classifier performance, they counted dimer occurrences from PSSM instead of counting it from the original protein sequence.

In this study, we developed a new definition of the RFAT using a different classification of amino acids. Based on the chemical property of the side chain of amino acids, we clustered 20 amino acids into four categories: {GAVLIMP}, {STCNQ}, {KRHED}, and {FYW}.[14] In this classification of amino acids, there are $4 \times 4 \times 4 = 64$ possible amino acid triplets. For each pair of virus and human proteins, we represent the RFAT as a 128-dimensional vector (64 for a virus protein and 64 for a human protein). The RFAT of the *i*th amino acid triplet is defined by Eq. (3).

$$\text{RFAT}_i = e^{\frac{f_i - \text{avg}\{f_1, f_2, \ldots, f_{64}\}}{\max\{f_1, f_2, \ldots, f_{64}\} - \text{avg}\{f_1, f_2, \ldots, f_{64}\}}}. \tag{3}$$

The minimum frequency of amino acid triplets in Cui's definition is replaced by the average frequency in our definition and 1 is not subtracted from the relative frequency. This definition reduces the number of zero elements in a feature vector, thus contributes to improve the performance of prediction models.

There are several differences between our representation and that of Cui *et al.*[8] First, we use a different definition of the RFAT using different classification of amino acids. This representation generates a feature vector with a much smaller number of zeros. Second, we encode virus protein sequences as well as human protein sequences in a feature vector, whereas Cui's representation encodes human protein sequences only and the type of a virus protein instead of virus protein sequences. Third, we use two additional features besides the RFAT, which are described below.

As an additional feature, we used the FDAT between virus and host proteins, defined by Eq. (4). In Eq. (4), f_{vi} is the frequency of amino acid triplet i in the virus protein of a protein pair, and f_{hi} the frequency of amino acid triplet i in the human protein of the same protein pair.

$$\text{FDAT}_i = e^{\frac{|f_{vi}-f_{hi}|-\text{avg}\{|f_{v1}-f_{h1}|,|f_{v2}-f_{h2}|,\ldots,|f_{v64}-f_{h64}|\}}{\max\{|f_{v1}-f_{h1}|,\ldots,|f_{v64}-f_{h64}|\}-\text{avg}\{|f_{v1}-f_{h1}|,\ldots,|f_{v64}-f_{h64}|\}}}.$$ (4)

We also encoded AC in a pair of virus and human proteins. AC is simply the frequency of each amino acid present in protein pairs divided by the maximum frequency of the amino acid in all protein pairs of a training dataset. The same maximum frequency of an amino acid obtained from a training dataset was used to encode the AC in protein pairs of a test dataset.

$$\text{AC}_i = \frac{f_i}{\max\{f_1, f_2, \ldots, f_{20}\}}.$$ (5)

Encoding all three features in a feature vector requires a total of 212 elements (128 for RFAT, 64 for FDAT, and 20 for AC) for each protein pair.

2.2. *Support Vector Machine*

We built several SVM models using LIBSVM[15] to predict the interactions between virus and human proteins using different combinations of the three features (RFAT, FDAT, and AC). The radial basis function (RBF) was used as a kernel function of the SVM models, and the best values of parameters C and γ of RBF were found from the grid parameter search of LIBSVM. Unless specified otherwise, the results of our SVM model for HPV–human PPIs were obtained with $C = 8$, $\gamma = 0.03125$, and those for HCV–human PPIs were obtained with $C = 8$, $\gamma = 0.03$.

2.3. *Datasets of HPV–Human PPIs and HCV–Human PPIs*

Interactions of human proteins with two types of viruses were used to evaluate the SVM models: HPV and HCV. HPV is a DNA virus from the papillomavirus family and HPV types 16 and 18 are known to be the most common cause of cervical cancer cases.[3,4] The HPV DNA encodes eight proteins (E1, E2, E4, E5, E6, E7, L1, and L2), which are required for viral DNA replication, encapsulation, and virus release.

We obtained the interaction data of HPV types 16 and 18 proteins with human proteins from the IntAct database (http://www.ebi.ac.uk/intact), and protein sequences from the Uniprot database (http://www.uniprot.org). Out of 20 different types of PPIs of IntAct, interactions annotated as 'physical association' or 'direct interactions' were used to construct a positive dataset, as in the study by Emamjomeh *et al.*[11] We removed human proteins with a sequence similarity higher than 80% using CD-HIT (http://cd-hit.org), and obtained 783 human proteins that interact with at least one of the eight HPV proteins. The positive dataset of HPV–human PPIs contains a total of 1031 interacting protein pairs.

SVM models require both positive and negative data, and negative data are not readily available in a database. We randomly selected 783 human proteins from 19,595 human proteins of Uniprot, which were not present in the positive dataset and constructed a negative dataset of 1031 protein pairs.

We divided both positive and negative datasets into training and test sets. We randomly selected 696 protein pairs (67.5% of the 1031 protein pairs) from each of the positive and negative datasets for a training dataset. The remaining 335 protein pairs in each of the positive and negative datasets were used to construct a test dataset of 335 positive data and 335 negative data. To maintain the same proportion of human proteins interacting with each virus protein, we selected training data by Eq. (6), as in the study by Cui *et al.*[8]

$$N_i = N(\text{Training}) \cdot \frac{N(T_i)}{N(\text{Total})}, \tag{6}$$

where T_i denotes the ith virus protein ($i = 1, 2, \ldots, 8$ for HPV proteins), $N(T_i)$ is the number of human proteins interacting with the ith virus protein, $N(\text{Training})$ is the total number of positive training data, and $N(\text{Total})$ is the total number of human proteins interacting with a virus protein. Table 1 shows the number of human proteins that are known to interact with each HPV protein and that selected for a training dataset.

The second type of viruses used to predict virus–human PPIs is HCV. HCV is a positive stranded RNA virus and belongs to the Flaviviridae family of genus Hepacivirus.[16] Approximately 170 million people (about 2% of world population) are infected by HCV. Except

Table 1. The number of human proteins interacting with HPV proteins.

HPV proteins	Total number of H_{HPV}	Number of H_{HPV} in a training set
E1	3	2
E2	60	40
E5	418	281
E6	197	132
E7	343	235
L1	2	1
L2	8	5
Total	1031	696

Note: H_{HPV} represents human proteins interacting with HPV proteins. For each HPV protein, the number of H_{HPV} in a training dataset was computed by $N_i = N(\mathrm{Training})\frac{N(T_i)}{N(\mathrm{Total})}$, where T_i is the ith HPV protein. Since some human proteins interact with multiple HPV proteins, the total numbers denote the number of interacting pairs of HPV and human proteins.

Table 2. The number of human proteins interacting with HCV proteins.

HCV proteins	Total number of H_{HCV}	Number of H_{HCV} in a training set
Core	102	71
E1	13	9
E2	25	17
F	21	15
NS2	10	7
NS3	235	164
NS4A	8	6
NS4B	4	3
NS5A	131	92
NS5B	33	23
P7	17	12
Total	599	419

Note: H_{HCV} represents human proteins interacting with HCV proteins. Since some human proteins interact with multiple HCV proteins, the total numbers denote the number of interacting pairs of HCV and human proteins.

retroviruses, HCV is the only RNA virus that is associated with liver cancer.[16] We found some redundancy in Cui's dataset of HCV–human PPIs,[8] and thus constructed new training and test datasets of HCV–human PPIs after removing redundant data. Table 2 shows the number of human proteins that are known to interact with each HCV protein, and that selected for a training dataset.

3. Results and Discussion

3.1. *Performance measures*

The performance of the prediction models were evaluated by six measures: sensitivity, specificity, accuracy, positive predictive value (PPV), negative predictive value (NPV), and Matthews correlation coefficient (MCC).

$$\text{Sensitivity} = \frac{TP}{TP + FN}, \tag{7}$$

$$\text{Specificity} = \frac{TN}{TN + FP}, \tag{8}$$

$$\text{Accuracy} = \frac{TP + TN}{TP + TN + FP + FN}, \tag{9}$$

$$\text{PPV} = \frac{TP}{TP + FP}, \tag{10}$$

$$\text{NPV} = \frac{TN}{TN + FN}, \tag{11}$$

$$\text{MCC} = \frac{TP \times TN - FP \times FN}{\sqrt{(TP + FP)(TP + FN)(TN + FP)(TN + FN)}}. \tag{12}$$

In Eqs. (7)–(12), true positives (TP) are human proteins that are correctly predicted as interacting with a virus protein. True negatives (TN) are non-interacting human proteins that are correctly predicted as non-interacting with a virus protein. False positives (FP) are non-interacting human proteins that are incorrectly predicted as interacting with a virus protein. False negatives (FN) are interacting human proteins that are incorrectly predicted as non-interacting with a virus protein.

3.2. *Interactions Between HPV and Human Proteins*

To examine the contribution of different features to the prediction performance, we built seven SVM models using different combinations of three features: the RFAT in virus and human proteins, FDAT between virus and human proteins, and AC in a pair of virus and human proteins.

Table 3. 10-fold cross-validation with datasets of HPV–human PPIs using different combinations of features.

Features	Sensitivity (%)	Specificity (%)	Accuracy (%)	PPV (%)	NPV (%)	MCC
RFAT	89.8	85.3	87.6	85.9	89.3	0.752
FDAT	91.9	86.3	89.1	87.0	91.5	0.783
AC	97.1	97.7	97.4	97.7	97.1	0.948
RFAT + FDAT	91.9	90.7	91.3	90.8	91.8	0.826
RFAT + AC	99.8	97.9	98.9	98.1	99.8	0.979
FDAT + AC	98.4	99.3	98.9	99.3	98.4	0.977
RFAT + FDAT + AC	**99.4**	**99.6**	**99.5**	**99.6**	**99.4**	**0.989**

Note: RFAT: relative frequency of amino acid triplets, FDAT: frequency difference of amino acid triplets between virus and host proteins, AC: amino acid composition, PPV: positive predictive value, NPV: negative predictive value, MCC: Matthews correlation coefficient.

We performed 10-fold cross-validations for each combination of features. Table 3 shows the results of 10-fold cross-validations with the data of HPV–human PPIs. All features showed a relatively good performance in all measures, but using all three features RFAT, FDAT, and AC resulted in a better performance (sensitivity = 99.4%, specificity = 99.6%, accuracy = 99.5%, PPV = 99.6%, NPV = 99.4%, and MCC = 0.989) than using a single or two features.

A typical cross-validation is known to lead to inflated accuracy values for paired inputs such as PPIs. In their recent works, Park and Marcotte[17] and Hamp and Rost[18] have demonstrated that both standard and refined cross-validations can lead to inflated accuracy of PPI prediction methods.

In this study, we performed leave one protein out (LOPO) cross-validations[19] as follows: (1) LOPO cross-validations with respect to virus proteins and (2) LOPO cross-validations with respect to human proteins. In LOPO cross-validations with respect to virus proteins, all protein pairs with one HPV virus protein are taken out for testing and remaining pairs are used for training. In LOPO cross-validations with respect to human proteins, all protein pairs with one human protein are taken out for testing and remaining pairs are used for training.

As shown in Table 4, the LOPO cross-validations of our method with respect to HPV proteins showed a lower performance (on average, sensitivity of 96.1%, specificity of 63.2%, accuracy of 79.7%, PPV of 72.3%, and NPV of 94.2%) than the 10-fold cross-validations in all performance measures, but the sensitivity and NPV of the LOPO cross-validations are still high.

Table 4. LOPO cross-validations of our method with respect to HPV proteins in the dataset of 696 HPV–human PPIs.

HPV	#human proteins (positives/ negatives)	TP	TN	FP	FN	SN	SP	AC	PPV	NPV
E1	2/2	2	0	2	0	100.00	0.00	50.00	50.00	0.00
E2	40/40	35	25	15	5	87.50	62.50	75.00	70.00	83.33
E5	281/281	280	110	171	1	99.64	39.15	69.40	62.08	99.10
E6	132/132	130	99	33	2	98.48	75.00	86.74	79.75	98.02
E7	235/235	220	201	34	15	93.62	85.53	89.57	86.61	93.06
L1	1/1	1	0	1	0	100.00	0.00	50.00	50.00	0.00
L2	5/5	1	5	0	4	20.00	100.00	60.00	100.00	55.56
Total	696/696	669	440	256	27	96.12	63.22	79.67	72.32	94.22

Note: SN: sensitivity, SP: specificity, AC: accuracy, PPV: positive predictive value, NPV: negative predictive value, MCC: Matthews correlation coefficient.

The result of LOPO cross-validations of our method with respect to human proteins cannot be shown in a table due to a large number of human proteins. On average, the LOPO cross-validations with respect to human proteins showed a much better performance than the LOPO cross-validations with respect to HPV proteins. On average, they achieved a sensitivity of 96.6%, a specificity of 96.7%, an accuracy of 96.6%, PPV of 96.7%, and NPV of 96.6% (TP: 672, TN: 673, FP: 23, FN: 24).

Due to the randomness in drawing negative data, we prepared additional five independent test datasets of HPV–human PPIs (Additional file 1). When selecting negative data, the maximum sequence similarity between positive and negative data and that within negative data were set to 80%.

To examine the effect of unbalanced positive and negative datasets on the prediction performance, we constructed following test datasets of three types:

(1) 5 datasets with 1:1 ratio of positive to negative instances.
(2) 5 datasets with 1:5 ratio of positive to negative instances.
(3) 5 datasets with 1:10 ratio of positive to negative instances.

Table 5 shows the results of testing our SVM model on the 15 test datasets. As expected, testing on unbalanced datasets resulted in lower performances than testing on balanced datasets with 1:1 ratio of positive to negative instances. ROC curves obtained from testing the SVM model on

Table 5. Results of independent testing of our SVM model on 15 different datasets of HPV–human PPIs (five datasets with 1:1 ratio of positive to negative instances, five datasets with 1:5 ratio, and five datasets with 1:10 ratio) using all three features (RFAT, FDAT, and AC). The SVM model used for the independent testing was generated with $C = 8$ and $\gamma = 0.03125$.

P:N	Test set	SN (%)	SP (%)	AC (%)	PPV (%)	NPV (%)	MCC
1:1	1	64.50	66.60	65.50	65.90	32.20	0.31
	2	64.50	69.60	67.01	67.92	66.20	0.34
	3	64.50	73.10	68.80	70.60	67.30	0.38
	4	64.50	62.99	63.70	63.50	63.90	0.28
	5	64.50	74.30	69.40	71.50	67.70	0.39
	Mean ± SD	64.50 ± 0.00	69.32 ± 4.65	66.88 ± 2.35	67.88 ± 3.30	59.46 ± 15.31	0.34 ± 0.05
1:5	1	64.50	66.70	66.40	27.94	90.40	0.24
	2	64.50	62.30	62.70	25.50	89.80	0.20
	3	64.50	64.40	64.40	26.60	90.10	0.22
	4	64.50	62.60	62.93	25.70	89.80	0.21
	5	64.50	66.70	66.40	27.94	90.40	0.24
	Mean ± SD	64.50 ± 0.00	64.54 ± 2.13	64.57 ± 1.80	26.74 ± 1.17	90.10 ± 0.30	0.22 ± 0.02
1:10	1	64.50	67.97	67.70	16.80	95.03	0.20
	2	64.50	64.70	64.70	15.50	94.80	0.17
	3	64.50	67.60	67.30	16.60	95.00	0.19
	4	64.50	62.80	62.96	14.80	94.70	0.16
	5	64.50	65.70	65.60	15.80	94.90	0.18
	Mean ± SD	64.50 ± 0.00	65.75 ± 2.13	65.65 ± 1.94	15.90± 0.82	94.89 ± 0.14	0.18 ± 0.02

Note: SN: sensitivity, SP: specificity, AC: accuracy, PPV: positive predictive value, NPV: negative predictive value, MCC: Matthews correlation coefficient.

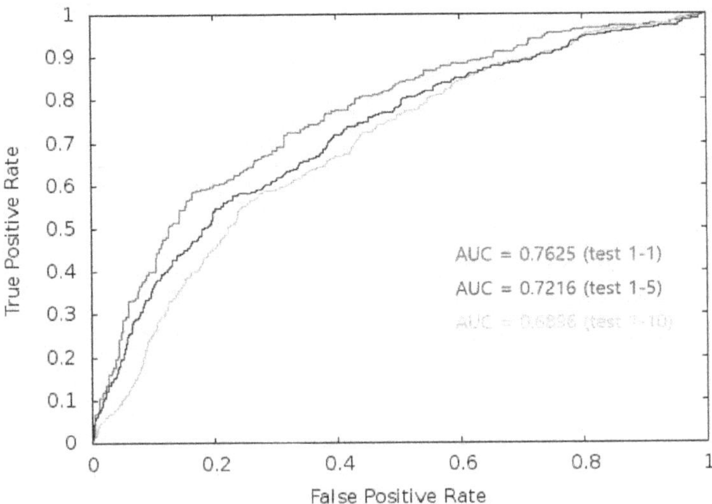

Fig. 1. ROC curves obtained from testing of our method on the datasets with different ratios of positive to negative instances of HPV–human PPIs.

the datasets of HPV–human PPIs are shown in Fig. 1. Testing the SVM model on the dataset with 1:1 ratio of positive to negative instances yielded the largest area under the ROC (AUC) value of 0.7625, whereas testing on the dataset with 1:10 ratio yielded the smallest AUC value of 0.6896. The 15 test datasets are available as Additional file 3.

To find new human proteins that potentially interact with HPV proteins, we searched HPRD (http://www.hprd.org/) for human proteins with the sequence similarity between 60% and 95% with any of 1032 known H_{HPV} proteins. After removing redundant sequences, we obtained a total of 357 human proteins as the initial candidates of H_{HPV} (Table 6).

Out of the 357 candidates, 101 human proteins were predicted as potential H_{HPV} by our SVM model. The 101 H_{HPV} candidates were refined by selecting human proteins that have the same GO cellular component[20] as known H_{HPV} for each HPV protein. After the refinement, we obtained a total of 50 candidates of H_{HPV}. For example, there are 60 H_{HPV} that are known to interact with the HPV E2 protein, and a total of 29 GO cellular component terms are annotated to the 60 H_{HPV}. There are 23 human proteins with the sequence similarity between 60% and 95% with 60 H_{HPV}. Our SVM model predicted 11 out of the 23 human proteins as interacting partners of the HPV E2 protein. 10 out of the 11 human proteins were left as reliable candidates of

Table 6. The new human proteins found by our method as potential interaction partners with HPV proteins.

HPV proteins	Number of H_{HPV} (GOs)	Initial candidates of H_{HPV} by sequence similarity	Predicted candidates by SVM (#GOs)	Refined candidates of H_{HPV} with GO
E1	3 (3)	2	1 (1)	1
E2	60 (29)	23	11 (3)	10
E4	1 (1)	0	0	0
E5	418 (59)	80	15 (4)	9
E6	197 (51)	92	16 (1)	8
E7	343 (77)	149	33 (5)	21
L1	2 (1)	6	5 (1)	1
L2	8 (4)	4	2 (2)	1
Total	1,032 (225)	357	101 (17)	50

Note: The 'Initial candidates of H_{HPV} by CD-HIT clustering' indicates the initial candidates of human proteins interacting with HPV proteins (called H_{HPV} in this paper) found by CD-HIT clustering with the known H_{HPV} as query sequences against human proteins from HPRD. The 'Predicted candidates of H_{HPV} by SVM' were determined by the SVM model from the initial candidates of H_{HPV}. The 'Refined candidates of H_{HPV} with GO' were obtained from the predicted candidates by selecting H_{HPV} that have the same GO cellular component terms as the known H_{HPV}.

H_{HPV} since they share at least one GO cellular component terms as the known H_{HPV} proteins. Figure 2 shows a PPI network of the 50 new H_{HPV} predicted by our method.

3.3. *Interactions Between HCV and Human Proteins*

We constructed SVM models for HCV–human PPIs in the same way as we did for HPV–human PPIs. We performed independent testing with different combinations of the three features, RFAT, FDAT, and AC. We prepared three independent test datasets of HCV–human PPIs by randomly selecting human proteins which are not present in the positive dataset of HCV–human PPIs with sequence similarity less than 80% (Additional file 2), as we did for HPV–human PPIs.

Table 7 shows the results of testing our SVM model on three test datasets using all three features RFAT, FDAT, and AC. The SVM model for HCV–human PPIs showed a reasonably high performance (on average, sensitivity=70.8%, specificity=79.6%, accuracy=75.2%, PPV=77.5%, NPV=73.3%, and MCC=0.506).

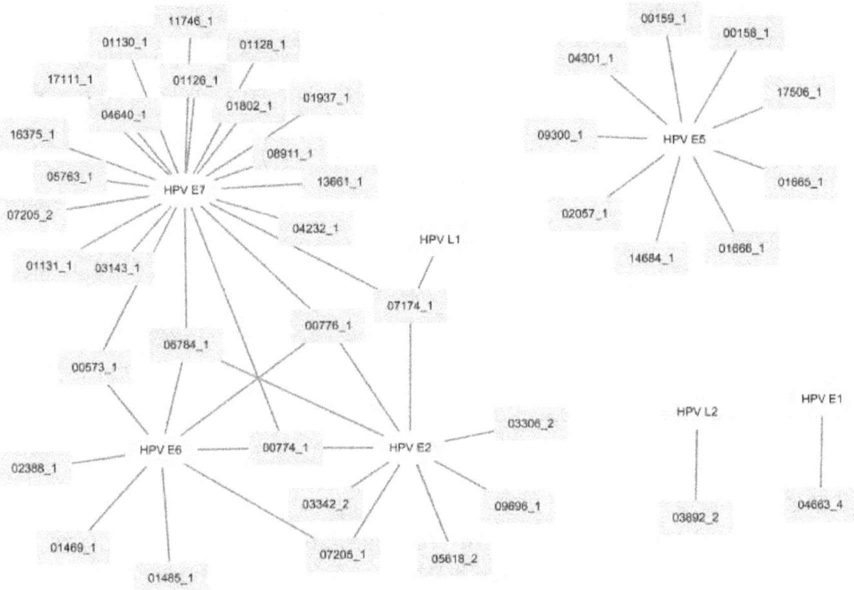

Fig. 2. A PPI network between HPV proteins and 50 new human proteins predicted by our method. HPV proteins are represented by yellow ellipses, and HPRD IDs of the human proteins are shown in blue rectangles.

Table 7. Results of independent testing of our method on three datasets of HCV–human PPIs. All three features (RFAT, FDAT, and AC) were used in testing.

Test set	Sensitivity (%)	Specificity (%)	Accuracy (%)	PPV (%)	NPV(%)	MCC
1	70.8	78.4	74.6	76.2	73.2	0.493
2	70.8	77.9	74.4	76.2	72.7	0.488
3	70.8	82.6	76.7	80.2	73.9	0.537
Average	70.8	79.6	75.2	77.5	73.3	0.506

Note: PPV: positive predictive value, NPV: negative predictive value, MCC: Matthews correlation coefficient. The three datasets of HCV-human PPIs are available in Additional file 2.

3.4. *Comparison to other Methods with Independent Datasets*

For comparative purposes, we tested our method, Cui's method[8] and Shen's method[12] on same datasets of HPV–human PPIs and HCV–human PPIs. Table 8 shows the results of testing the three methods on three test datasets of HPV–human PPIs. The three methods used SVM models with

Table 8. Comparison of three methods on three test datasets of HPV-human PPIs.

	Test set	Sensitivity (%)	Specificity (%)	Accuracy (%)	PPV (%)	NPV (%)	MCC
Our method	1	64.5	66.6	65.5	65.9	65.2	0.311
($c = 8$, $\gamma = 0.03125$)	2	64.5	69.6	67.0	67.9	66.2	0.341
	3	64.5	73.1	68.8	70.6	67.3	0.378
Cui's method	1	41.5	77.3	59.4	64.7	56.9	0.201
($c = 8$, $\gamma = 0.5$)	2	41.5	74.3	57.9	61.8	56.0	0.167
	3	41.5	77.9	59.7	65.2	57.1	0.208
Shen's method	1	56.7	57.9	57.3	57.4	57.2	0.146
($c = 8$, $\gamma = 0.125$)	2	56.7	59.4	58.1	58.3	57.8	0.161
	3	56.7	67.2	61.9	63.3	60.8	0.240

Note: PPV: positive predictive value, NPV: negative predictive value, MCC: Matthews correlation coefficient.

RBF as their kernel. For a fair comparison, the best values of the parameters C and γ of the three SVM models were obtained separately with the training dataset of HPV–human PPIs, and the values were used in testing the models. Nonetheless, our method showed a much better performance than the other two methods. ROC curves obtained from testing the three prediction methods on the datasets of HPV–human PPIs are shown in Fig. 3. Our method showed the largest AUC value of 0.7625.

Table 9 shows the results of testing the three methods on three test datasets of HCV–human PPIs. Both Cui's method and Shen's method showed a better performance with the datasets of HCV–human PPIs than with the datasets of HPV–human PPIs. However, our method still shows the best performance in all performance measures with the datasets of HCV–human PPIs. The results of testing three methods on the datasets of HPV–human PPIs and HCV–human PPIs demonstrate that our method is much better than the others in all performance measures.

A recent method developed by Emamjomeh *et al.*[11] showed a sensitivity of 84%, specificity of 87%, and accuracy of 83% with the dataset of HCV–human PPIs. But, their performance is lower than the average performance of our method with three datasets of HCV–human PPIs (average sensitivity of 89.4%, specificity=88.1%, accuracy=88.8%, PPV=88.6%, NPV=88.8%, and MCC=0.774 in Table 9). Another recent work by Barman *et al.*[9] showed an accuracy of 80% with the dataset of HCV–human PPIs, which is lower than the average accuracy of 88.8% of our method with three datasets of HCV–human PPIs.

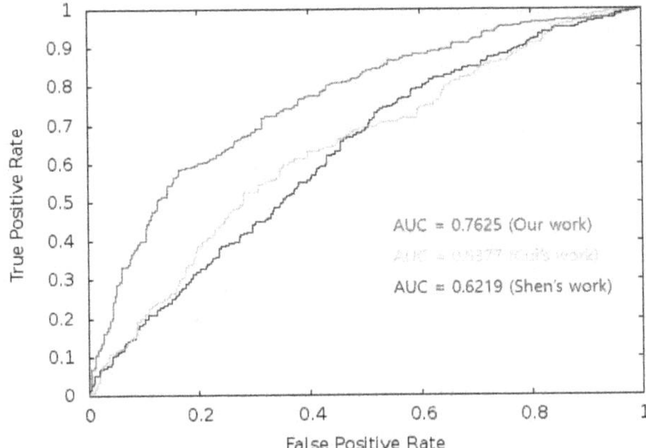

Fig. 3. ROC curves obtained from testing three prediction methods (our method, Cui's method, and Shen's method) on same test datasets with 1:1 ratio of positive to negative instances of HPV–human PPIs.

Table 9. Comparison of three methods on three independent datasets of HCV–human PPIs. The three datasets of HCV-human PPIs are available in Additional file 2.

	Test set	Sensitivity (%)	Specificity (%)	Accuracy (%)	PPV (%)	NPV (%)	MCC
Our method	1	70.8	78.4	74.6	76.2	73.2	0.493
($c = 8$, $\gamma = 0.03$)	2	70.8	77.9	74.4	76.2	72.7	0.488
	3	70.8	82.6	76.7	80.2	73.9	0.537
Cui's method	1	64.3	78.2	71.3	77.3	65.2	0.414
($c = 8$, $\gamma = 0.5$)	2	64.2	77.9	71.1	76.1	66.1	0.427
	3	64.4	79.4	71.9	78.1	65.8	0.434
Shen's method	1	70.1	60.8	65.4	61.4	69.4	0.381
($c = 8$, $\gamma = 0.125$)	2	70.2	62.2	66.2	62.1	70.2	0.396
	3	70.1	61.3	65.7	61.9	69.5	0.384

Note: PPV: positive predictive value, NPV: negative predictive value, MCC: Matthews correlation coefficient.

4. Conclusions

So far most computational methods developed for predicting PPIs are intended for interactions within a species rather than for interactions across different species such as interactions between virus and host cell proteins. In this study, we improved the previous representation methods and identified powerful features for predicting the interactions between virus and human proteins. We represented the RFAT in virus and human proteins, FDAT between virus and human proteins, and AC in a pair of virus and human proteins in a feature vector. AC is a commonly used in predicting PPIs. But, RFAT and FDAT between virus and host proteins were used in our study for the first time to predict the interactions between virus and host proteins.

We built SVM models to predict human proteins that interact with HPV proteins and HCV proteins. The SVM model for predicting HPV–human PPIs showed an average accuracy of 66.9% in five independent datasets, which were not used in training the models with sequence similarity less than 80%, whereas the SVM model for predicting HCV–human PPIs showed an average accuracy of 75.2%, in three independent datasets.

Comparison of our SVM models to other SVM models with the same datasets demonstrated that our models are better than the others in all performance measures. Using the new features and representation method, we represented a large amount of data involved in PPIs in feature vectors of small and fixed size, and achieved a substantially higher performance than previous computational methods. The features and representation method will be useful in predicting other types of heterogeneous PPIs.

Acknowledgments

This research was supported by the international cooperation program managed by the National Research Foundation (NRF) (2014K2A2A2000670) and in part by Basic Science Research Program through NRF funded by the Ministry of Science, ICT & Future Planning (2015R1A1A3A04001243).

References

1. Navratil V, de Chassey B, Meyniel L, Delmotte S, Gautier C, André P, Lotteau V, Rabourdin-Combe C, VirHostNet: A knowledge base for the management and the analysis of proteome-wide virus–host interaction networks, *Nucl Acids Res* **37**(1):661–668, 2009.
2. Zhou H, Jin J, Wong L, Progress in computational studies of host-pathogen interactions, *J Bioinform Comput Biol* **11**(2):1230001, 2013.

3. McLaughlin-Drubin ME, Munger K, Biochemical and functional interactions of human papillomavirus proteins with polycomb group proteins, *Viruses* **5**(3):1231–1249, 2013.

4. Facciuto F, Valdano MB, Marziali F, Massimi P, Banks L, Cavatorta AL, Gardiol D, Human papillomavirus (HPV)-18 E6 oncoprotein interferes with the epithelial cell polarity Par3 protein, *Mol Oncol* **8**(3):553–543, 2014.

5. Mukhopadhyay A, Ray S, Maulik U, Incorporating the type and direction information in predicting novel regulatory interactions between HIV-1 and human proteins using a biclustering approach, *BMC Bioinformatics* **15**(26), 2014.

6. Zhou H, Rezaei J, Hugo W, Gao S, Jin J, Fan M, Yong C, Wozniak M, Wong L, Stringent DDI-based prediction of H. sapiens-M. tuberculosis H37Rv protein-protein interactions, *BMC Syst Biol* **7**(Suppl 6):S6, 2013.

7. Zhou H, Gao S, Nguyen NN, Fan M, Jin J, Liu B, Zhao L, Xiong G, Tan M, Li S, Wong L, Stringent homology-based prediction of H. sapiens-M. tuberculosis H37Rv protein-protein interactions, *Biol Direct* **9**:1–30, 2014.

8. Cui G, Fang C, Han K, Prediction of protein–protein interactions between viruses and human by an SVM model, *BMC Bioinformatics* **13**(7):S5, 2012.

9. Barman RK, Saha S, Das S, Prediction of interactions between viral and host proteins using supervised machine learning methods, *PLoS ONE* **9**(11): e112034, 2014.

10. Chatr-aryamontri A, Ceol A, Peluso D, Nardozza A, Panni S, VirusMINT: A viral protein interaction database, *Nucl Acids Res* **37**:669–673, 2009.

11. Emamjomeh A, Goliaei B, Zahiri J, Ebrahimpour R, Predicting protein–protein interactions between human and hepatitis C virus via an ensemble learning method, *Mol Biosyst* **10**(12):3147–3154, 2014.

12. Shen J, Zhang J, Luo X, Zhu W, Yu K, Li Y, Jiang H, Predicting protein-protein interactions based only on sequences information, *Proc Natl Acad Sci USA* **104**:4337–4341, 2007.

13. Sharma A, Lyons J, Dehzangi A, Paliwal KK, A feature extraction technique using bi-gram probabilities of position specific scoring matrix for protein fold recognition, *J Theor Biol* **320**:41–46, 2013.

14. Dong Y, Kuang Q, Dai X, Li R, Wu Y, Leng W, Li Y, Li M, Improving the understanding of pathogenesis of human papillomavirus 16 via mapping protein-protein interaction network, *Biomed Res Intl* **2015**:890381, 2015.

15. Chang C, Lin C, LIBSVM: A library for support vector machines, *ACM Trans Intell Syst Technol* **2**(3):27, 2011.

16. Ramareddy VG, Mythili KP, Interaction of Hepatitis C viral proteins with cellular oncoproteins in the induction of liver cancer, *ISRN Virol* **2014**:351407, 2014.

17. Park Y, Marcotte EM, A flaw in the typical evaluation scheme for pair-input computational predictions, *Nat Methods* **9**(12):1134–1136. 2012.

18. Hamp T, Rost B, More challenges for machine-learning protein interactions, *Bioinformatics* **31**(10):1521–1525, 2015.
19. Abbasi WA, Minhas FA, Issues in performance evaluation for host–pathogen protein interaction prediction, *J Bioinform Comput Biol*, 2016, DOI: 10.1142/S0219720016500116.
20. Ashburner M, Ball CA, Blake JA, Botstein D, Butler H, Cherry JM, Davis AP, Dolinski K, Dwight SS, Eppig JT, Harris MA, Hill DP, Issel-Tarver L, Kasarskis A, Lewis S, Matese JC, Richardson JE, Ringwald M, Rubin GM, Sherlock G, Gene ontology: Tool for the unification of biology, *Nat Genet* **25**(1):25–29, 2000.
21. You Z-H, Chan KC, Hu P, Predicting protein-protein interactions from primary protein sequences using a novel multi-scale local feature representation scheme and the random forest, *PLoS ONE* **10**(3):e0125811, 2015.

Byungmin Kim received a BS in Information and Communication Engineering, Inha University in 2008 and an MS in Computer Science and Engineering, Inha University, Korea in 2016.

Saud Alguwaizani is a PhD student at the Department of Computer Science and Engineering, Inha University Incheon, Korea. He received a BS in Computer and Information System in 2005 from King Faisal University, Alhasa, Saudi Arabia. He received an MS in Information Technology Management in 2009 from Wollongong University, Wollongong, Australia. His research areas include bioinformatics, AI and image processing.

Xiang Zhou received his Bachelor degree from Anhui University of Traditional Chinese Medicine, China and Soonchunhyang University, Korea, in 2012 and 2014, respectively. He is currently pursuing master's degree in Computer Science and Engineering, Inha University, Korea. His research interests include bioinformatics and data mining.

De-Shuang Huang received B.Sc., M.Sc. and Ph.D. degrees in electronic engineering from Institute of Electronic Engineering, Hefei, National Defense University of Science and Technology, Changsha, and Xidian University, Xian, China, in 1986, 1989 and 1993, respectively. During 1993–1997 he was a postdoctoral researcher in Beijing Institute of Technology and National Key Laboratory of Pattern Recognition and Chinese Academy of Sciences, Beijing, China. In September, 2000, he joined the Institute of Intelligent Machines, Chinese Academy of Sciences, Hefei, China as the recipient of "Hundred Talents Program of CAS". He is currently the head of Institute of Machines Learning and Systems Biology, Tongji University. His current research interest includes bioinformatics, pattern recognition and neural networks.

Byungkyu Park is a research professor at the Department of Computer Science and Engineering, Inha University, Incheon, Korea. He received a BS from Mathematics, Inha University in 2001, an MS in Computer Science and Engineering, Inha University in 2004, and a PhD in Computer Science and Information Engineering, Inha University in 2011. His research areas include bioinformatics and learning.

Kyungsook Han is a professor at the Department of Computer Science and Engineering, Inha University, Incheon, Korea. She received a BS cum laude from Seoul National University in 1983, an MS cum laude in Computer Science from Korea Advanced Institute of Science and Technology (KAIST) in 1985, another MS in Computer Science from the University of Minnesota at Minneapolis, USA in 1989, and a PhD in Computer Science from Rutgers University, USA in 1994. Her research areas include bioinformatics, visualization, and learning.

CHAPTER 8

Issues in Performance Evaluation for Host–Pathogen Protein Interaction Prediction[a]

Wajid Arshad Abbasi[*] and
Fayyaz Ul Amir Afsar Minhas[†,‡]

Department of Computer and Information Sciences
Pakistan Institute of Engineering and Applied Sciences (PIEAS)
Nilore, Islamabad, Pakistan
**wajidarshad@gmail.com*
†afsar@pieas.edu.pk

The study of interactions between host and pathogen proteins is important for understanding the underlying mechanisms of infectious diseases and for developing novel therapeutic solutions. Wet-lab techniques for detecting protein–protein interactions (PPIs) can benefit from computational predictions. Machine learning is one of the computational approaches that can assist biologists by predicting promising PPIs. A number of machine learning based methods for predicting host–pathogen interactions (HPI) have been proposed in the literature. The techniques used for assessing the accuracy of such predictors are of critical importance in this domain. In this paper, we question the effectiveness of K-fold cross-validation for estimating the generalization ability of HPI prediction for proteins with no known interactions. K-fold cross-validation does not model this scenario, and we demonstrate a sizable difference between its performance and the performance of an alternative evaluation scheme

---

[a]This article was previously published in *Journal of Bioinformatics and Computational Biology*. Vol: 14, No. 3 (2016), 1650011 (17 pages).

[‡]Corresponding author.

called leave one pathogen protein out (LOPO) cross-validation. LOPO is more effective in modeling the real world use of HPI predictors, specifically for cases in which no information about the interacting partners of a pathogen protein is available during training. We also point out that currently used metrics such as areas under the precision-recall or receiver operating characteristic curves are not intuitive to biologists and propose simpler and more directly interpretable metrics for this purpose.

Keywords: Performance evaluation; host–pathogen interactions; protein–protein interactions; machine learning; cross-validation.

1. Introduction

Invasion of a host organism by pathogens like viruses or bacteria and the reactions of the host to these pathogens is known as infection.[1] According to the World Health Organization (WHO), infectious diseases, such as Tuberculosis, Hepatitis, AIDS, and Typhoid etc., are among the major causes of deaths in the world. It is estimated that infections are involved in approximately 20% of all deaths across the globe.[2] Infection occurs through interactions between proteins of the host and the pathogen.[3] Therefore, to understand the underlying mechanism of infectious diseases, it is crucial to gain an insight into host–pathogen protein–protein interactions (PPIs).[4]

Experimental methods for studying PPIs are often time-consuming and expensive, making it difficult to investigate all possible host–pathogen interactions (HPI). For instance, the bacterium *Bacillus anthracis* has 5,508 protein encoding genes,[5] which when paired with the 20,000 or so human genes,[6] gives more than 100 million possible pairs to test experimentally. It is not practically feasible to test all possible interactions experimentally. Therefore, there is an utmost need for computational approaches to support wet-lab methods by predicting promising PPIs. Such computational approaches can assist biologists in focusing on the most likely interactions.[7]

Machine learning techniques for prediction of PPIs have been studied extensively in the bioinformatics community.[7–13] The general framework that describes the application of machine learning techniques in HPI prediction is shown in Fig. 1. This framework shows how the classifier is trained on training data and consequently used to predict the label of a novel example. In such methods, a pair of proteins, one from the host and one from the pathogen, is considered as a classification example. In machine learning

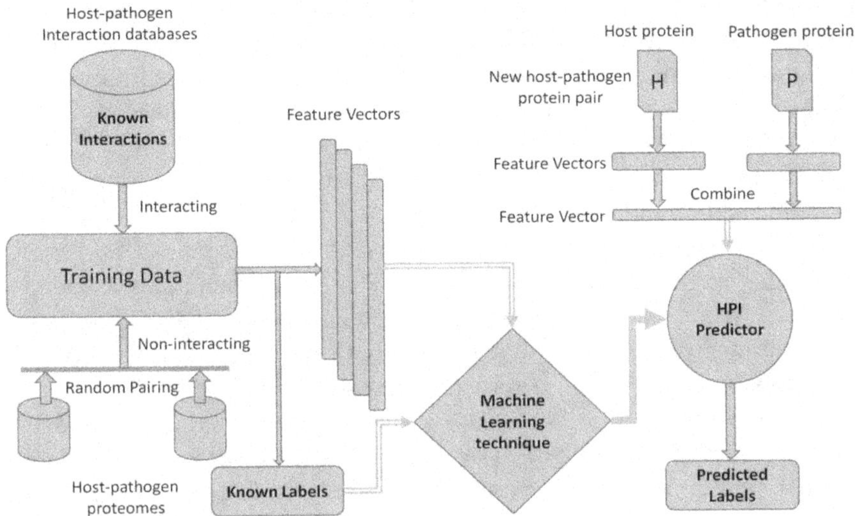

Fig. 1. Framework of machine learning techniques in host–pathogen PPIs prediction.

based HPI predictors, experimentally discovered interactions are used as positive examples in training.[7] Negative or noninteracting examples are usually generated by pairing host–pathogen proteins randomly.[14] To produce good classifiers, a considerable number of interacting and noninteracting pairs are typically needed.

For HPI predictors, predictive features are derived for each example using various attributes of the two proteins such as the k-mer composition of their protein sequences,[15,16] protein domain information,[3] gene expression,[11] Gene Ontology,[5,10] network characteristics,[5,9] evolutionary profiles,[17,18] etc.

There is a pressing need for accurate HPI predictors. A number of such predictors exist in the literature.[9–11,19–22] Our primary focus in this paper is to analyze the methods used for assessment of accuracy of HPI predictors. In every machine learning setting, it is important to consider the nature of the problem, available data and intended use of the system while designing the classifier, its evaluation protocols and performance metrics.[23,24] However, in comparison to other application areas, this requirement is even more crucial in bioinformatics because of its role as a tool for biological discovery. Specifically, in the domain of host–pathogen interaction prediction, it is important to be cognizant of the underlying biological implications in designing the machine learning system and its evaluation protocol. A number of factors affect the learning of an HPI predictor such as imbalanced data, high dimensionality of the feature space, sparseness of known interactions and potential uncertainties in labels of negative training examples.[14] A bias that affects the measurement of the generalization performance of an HPI predictor stems from the fact that, in such predictors, classification examples are pairs of proteins. Machine learning problems involving paired inputs lead to unique challenges in the assessment of accuracy of such techniques as pointed out by the recent papers of Park *et al.*[25] and Hamp *et al.*[26] In this work, we demonstrate the effect of this bias on performance assessment of HPI predictors through K-fold cross-validation by simulation studies.

Most existing HPI predictors use a very simple K-fold cross-validation (CV)[27] scheme for performance evaluation. In K-fold cross-validation, the original dataset is randomly partitioned into K equal sized subsets. Of the K subsets, K-1 sets are used for training and the remaining data is used in testing. This process is repeated K times to compute fold-wise performance measures which are then combined to produce a single accuracy metric.[27] This approach works very well in general, however, for HPI prediction a more elaborate analysis is required. This is because simple K-fold cross-validation does not take the biological nature of the problem into account. It does not prevent both similarity and redundancy between training and testing examples within a fold. A protein in a test example in a fold can occur as part of another example in training. This issue is depicted in Fig. 2 using a toy dataset, and can lead to inflated accuracy values for paired input problems as pointed out by Park and Marcotte.[25] Typically, a biologist is interested in finding interactions between proteins which may be very dissimilar in their sequence from the proteins used in training. However, in simple K-fold cross-validation, close sequence homologs of a test protein pair can appear in training if the dataset has not been made nonredundant. This issue was also observed in a study of intra-species PPIs by Hamp and Rost.[26] We hypothesize that this produces a disparity between the measured accuracy of an HPI predictor and its true generalization performance.

In this work, we test this hypothesis rigorously by carefully performing a number of simulation experiments with different performance assessment schemes. We also propose a cross-validation scheme specifically tailored for HPI predictors called Leave One-pathogen Protein Out (LOPO) cross-validation. LOPO cross-validation minimizes the

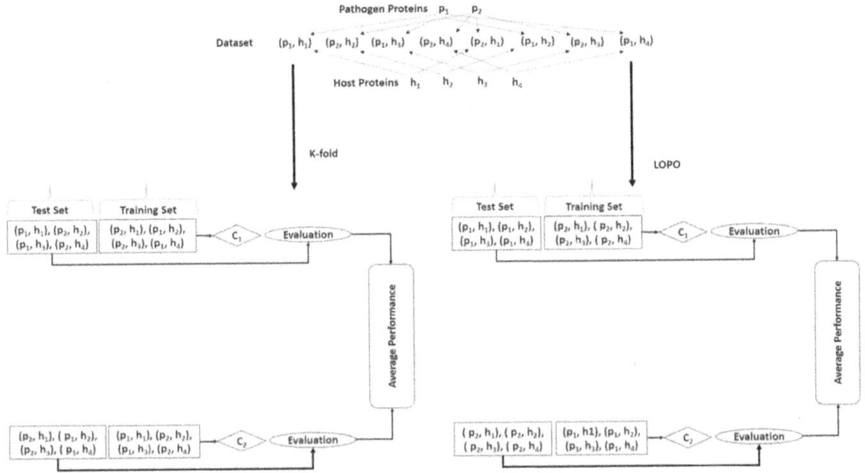

Fig. 2. Evaluation methodologies for HPI predictors. In K-fold cross-validation (shown on the left), we have separated our data into as many folds as the number of pathogen proteins. Then, one fold is taken for testing and all other folds are used for training. Redundancy with respect to both host and pathogen proteins can be seen. For example, proteins p_1 and h_1 occur in both training and test sets in each fold. In LOPO cross-validation, protein pairs with respect to one pathogen protein involved in interactions are taken out for testing and remaining pairs are used for training. Number of folds in this case will be equal to the number of pathogen proteins involved in interactions. Redundancy with respect to pathogen protein is eliminated in this evaluation protocol.

overlap between training and testing proteins in a fold. This leads to more realistic estimates of generalization performance for pathogen proteins for which no interactions are available during training.

We also point out another issue in the evaluation of HPI predictors. Accuracy measures, such as areas under the precision-recall (AUC-PR)[28] or receiver operating characteristic (AUC-ROC)[28] curves, accuracy and F1 score[11] are typically used in the area of machine learning and for presenting the results for an HPI predictor. These measures, though important in the analysis of classifiers, are not directly useful to a biologist who is interested in employing an HPI predictor for designing wet-lab experiments to identify potential interactions.[29] Therefore, we propose three new domain specific metrics in this paper called True Hit Rate (THR), False Hit Rate (FHR) and Median Rank of the First Positive Prediction (MRFPP) which can assist a biologist in experiment design.

2. Methods

In this section, we discuss the details of our experimental design to test our hypothesis about the overestimation of generalization performance as a consequence of K-fold cross-validation.

2.1. *Datasets and Preprocessing*

We have used two separate HPI datasets to see if the validity of our hypothesis remains unaffected by the choice of the data. The details of these datasets are given below.

2.1.1. *Human–HIV Interaction Dataset (HH)*

We extracted 632 known unique interactions of human proteins with proteins from the Human Immunodeficiency Virus (HIV, isolate HXB2 group M subtype B) from the NCBI HIV-human protein interaction database.[30] This set covers 513 human and 17 HIV proteins.

2.1.2. *Human–Adenovirus Interaction Dataset (HA)*

We used the Pathogen–Host Interaction Search Tool (PHISTO)[31] to develop a dataset of 212 known interactions of human and Adenovirus 5 proteins. This dataset contains 194 human and 9 Adenovirus proteins.

The proteins in both datasets are nonredundant with less than 80% sequence identity between them. This nonredundancy is required to minimize the impact of sequence homology on the outcome of our simulation experiments and was ensured using the CD-HIT[32] tool.

2.1.3. *Generation of Negative Examples*

We generated negative examples by pairing all the human and pathogen proteins in the above datasets and removing examples present in the positive set.[14] The total numbers of negative examples for the two datasets thus obtained are 8,089 and 1,534, respectively. It must be noted here that we purposefully did not take the whole human proteome in consideration as it would have complicated our evaluation. Our negative set includes proteins which are already involved in an interaction. This presents a harder classification problem in comparison to randomly selecting proteins from the proteomes of the two species for generating negative examples.

As a consequence, our results can be expected to be lower bounds of classification performance for proteome-wide evaluation.

2.2. *Classifiers*

In order to test whether the validity of our hypothesis remains unaffected by the choice of the classification scheme, we experimented with two different classifiers: A linear Support Vector Machines (SVM)[33] and a Random Forest classifier (RF).[34] We chose these classifiers because both of them have been widely used in the literature for the development of HPI predictors.[9,10,22,25,29,35–37] A brief description of both methods is given below. However, we begin by presenting HPI prediction as a classification problem. In machine learning based HPI prediction, a dataset S consisting of labeled examples $(x_i, y_i) \in S$ is given. Here, x_i is a feature vector for the ith example involving human protein h and pathogen protein p that form the pair (h, p). The corresponding label of each example is given by $y_i \in \{+1, -1\}$. A label of $y_i = +1$ indicates that the human protein h is known to interact with the pathogen protein p whereas $y_i = -1$ indicates a noninteracting protein pair. At an abstract level, the objective of classification for HPI prediction is to find a function $f(x)$ whose value is indicative of whether the given example, represented by x, is interacting or noninteracting. The classifier must generalize well over previously unseen data.

2.2.1. *SVMs*

SVMs are based on the principle of structural risk minimization[33] and offers a number of advantages including optimal margin classification, strong theoretical foundations and high generalization capabilities especially in high dimensional feature spaces. Training an SVM for HPI prediction involves optimizing the objective criteria given in Eq. (1) to find a linear discriminant function $f(x_i) = w^T x_i + b$ parameterized by a weight vector w and bias b.[33]

$$min_{w,b,\xi} \frac{1}{2} w^T w + C \sum_{i=1}^{|S|} \xi_i, \tag{1}$$

Subject to:

$$y_i(w^T x_i + b) \geq 1 - \xi_i, \xi_i \geq 0, \forall i.$$

Here, $\frac{1}{2}w^T w$ controls the inverse of the margin, ξ_i is the extent of margin violation for a given training example and C is the penalty of such violations.[33] The value of C was coarsely optimized during model selection. In order to avoid class imbalance, we chose the value of C for examples of each class to be proportional to the number of training examples of that class. We use the Scikit-learn[38] package for SVM training and testing.

2.2.2. *Random Forest (RF)*

RF[34] is an ensemble of decision trees based on randomly sampled subsets of input features. We coarsely optimized RF classifiers with respect to the number of decision trees and their maximum depth. The labels produced by individual trees in the forest are converted to probability values and used as the scoring function for a test example. To train and test the RF, we have used Scikit-learn.[38]

2.3. *Feature Extraction*

In this work, we have employed a number of feature representations to observe the impact of various features on the validity of our hypothesis. Features $\phi(\cdot)$ are extracted for individual proteins in a given example (h, p) and are concatenated to form features at the pair level as follows: $x = [\phi_1(h), \phi_2(h), \ldots, \phi_n(h), \phi_1(p), \phi_2(p), \ldots, \phi_n(p)]^T$. The individual protein level features are normalized to unit norm, i.e. $\|\phi(\cdot)\|_2 = 1.0$. In what follows, we describe the sequence-based protein-level feature representations used in this study.

2.3.1. *k-mer Composition*

In order to capture the sequence properties of a given protein sequence, we have used k-mer composition as a feature.[39] Protein sequence k-mer composition features are extracted by grouping the twenty naturally occurring amino acids into seven groups on the basis of their hydrophobic and electrostatic properties.[15] We used the counts of occurrences of k-mers of these groups in the protein sequence as features with value of k ranging from 2 to 4. In this way, each protein is represented in the form of feature vector of length 7^k, i.e. 49, 343, and 2401 for $k = 2, 3, 4$, respectively.

2.3.2. *ProFET Features (ProFET)*

To capture biophysical properties of amino acids in a given protein, we used the ProFET.[40] ProFET gives a 167 dimensional representation of a protein with features modeling the amino acid level entropy, autocorrelation, potential disorder and presence of significant motifs.

2.3.3. *BLOSUM Features (Blosum)*

In an effort to model the substitutions of physiochemically similar amino acids in proteins, we chose to represent each protein using the BLOSUM-62[41] substitution matrix. In this feature representation, a given protein p of length L is represented by a 20-dimensional feature vector, $\phi(p) = \frac{1}{L} \sum_{i=1}^{L} B(p_i)$, where $B(p_i)$ is the column in the BLOSUM-62 substitution matrix corresponding to the ith residue in p. Variants of these features have previously been used in related problems.[42–44]

2.3.4. *Evolutionary Features*

In order to model evolutionary relationships between proteins, we developed a feature representation based on the Position Specific Frequency Matrix (PSFM)[45] of a given protein. The PSFM is obtained by running PSI-BLAST[45] for the given protein for three iterations against the nonredundant (nr) protein database[46] with an E-value threshold of 10^{-3}. The PSFM for a protein p of length L is a $20 \times L$ dimensional matrix, F^p, with each column, F_i^p, representing the relative frequency of occurrence of different amino acids at position i in the homologous proteins returned by PSI-BLAST. The PSFM is then marginalized for position to produce a 20-dimensional feature representation $\phi(p) = \frac{1}{L} \sum_{i=1}^{L} F_i^p$. A similar representation has also been used in previous studies.[17,18]

2.4. *Model Evaluation*

In order to evaluate the performance of various machine learning models, we have used classical K-fold cross-validation for each dataset. As discussed earlier, our hypothesis hinges on the fact that it is possible for a protein in a testing example to occur as part of the training examples of a fold in K-fold cross-validation. To test this hypothesis, we compare K-fold cross-validation results with a LOPO cross-validation scheme which minimizes such overlaps between training and testing data within a fold. Henceforth, we describe each model evaluation scheme in more detail.

2.4.1. *K-fold Cross-Validation*

In order to keep the results of K-fold[27] and LOPO cross-validation schemes comparable, we chose the value of K in standard cross-validation to be equal to the number of pathogen proteins in the evaluation dataset. This corresponds to $K = 17$ and $K = 9$ for the Human-HIV and Human-Adenovirus datasets, respectively.

2.4.2. *LOPO*

In a single fold of LOPO cross-validation, we used all examples involving a pathogen protein in testing while all the remaining examples are used for training. This process is repeated for each pathogen protein in the given dataset. This concept is depicted in Fig. 2. As all the protein pairs belonging to a specific pathogen protein are taken out for testing, any overlap over that pathogen protein in training and testing is avoided. This evaluation protocol is related to the C2 cross-validation approach discussed for analysis of paired data by Park and Marcotte.[25] However, as opposed to the C2 cross-validation protocol, LOPO is centered on minimizing redundancy with respect to the pathogen protein only. A similar approach can be devised by making folds by holding-out all examples in the dataset that involve a host protein. In our datasets, the number of pathogen proteins is significantly smaller than those from the host and each pathogen protein interacts with numerous host proteins. On the other hand, a host protein typically interacts with a single pathogen protein only. As a consequence, LOPO cross-validation is stricter in terms of controlling the redundancy between training and test data.

2.5. *Performance Metrics*

We used the following metrics to evaluate and compare the performance of the trained models. These metrics are widely used in the evaluation of numerous HPI predictors. Using these two metrics, we can also test whether our hypothesis holds independent of the choice of the performance metric or not. These metrics are evaluated for each fold and their mean and standard deviation values across all folds in cross-validation are reported. The standard deviation gives an idea of the expected degree of variability in the results.

2.5.1. *Area Under the ROC Curve (AUC-ROC)*

The AUC-ROC curve is obtained by plotting the true positive rate versus the false positive rate at different thresholds on the discriminant function scores produced by the classifier.[28] An ideal predictor will score an AUC-ROC of 1.0 whereas random guessing will have a score of 0.5.

2.5.2. *Area Under the Precision-Recall Curve (AUC-PR)*

The scores produced by the predictor and the known labels for each example were used to plot precision-recall curves.[28] The area under the precision-recall curve (AUC-PR) expressed as a percentage, has also been used as the performance metric. Unlike AUC-ROC, this metric is sensitive to the number of false positives.

2.5.3. *Proposed Performance Metrics*

In this work, we propose the use of the following domain-specific metrics for host-pathogen PPIs predictors. In comparison to AUC-ROC and AUC-PR scores, these proposed metrics can be more helpful to a biologist in the design of wet-lab experiments to test the top scoring predictions of an HPI predictor. These metrics have previously been used in our work on identifying interfaces and binding sites in proteins.[23–25]

2.5.3.1. True Hit Rate (THR)

Intuitively, the THR tells a biologist how often the top scoring example involving a pathogen protein can be expected to be a true positive. When testing the examples involving a held out pathogen protein in LOPO cross-validation, we define a true hit to occur when the highest scoring example, amongst all the given examples, is a true positive. The percentage of true hits across all interacting pathogen proteins is called the true hit rate. For an ideal classifier, $THR = 100\%$.

2.5.3.2. False Hit Rate (FHR)

This metric tells the biologist the expected number of known negatives involving a protein that score higher than the top scoring true positive example. For a held out pathogen protein in LOPO cross-validation, this metric represents the percentage of negative examples involving that pathogen protein that score higher than its top scoring positive example.

For an ideal classifier, $FHR = 0\%$. This can be very useful in experimental design to screen for potential interaction partners of a protein in the wet-lab.

2.5.3.3. Median Rank of the First Positive Prediction (MRFPP)

This metric gives an intuitive idea of the distribution of false negatives in comparison to the top scoring true positive. This metric is simply the median rank of the first positive prediction across all pathogen proteins. An ideal predictor should have $MRFPP = 1$, i.e. for at least 50% pathogen proteins, the top scoring prediction by the predictor is a true positive.[35] In comparison to AUC-PR, this measure is more intuitive to biologist as it reveals directly how often the top scoring predictions can be expected to be an interacting pair.

It must be noted that these metrics give a general idea of the performance of the classifier. One possible shortcoming with these metrics is that they can possibly be affected by differences in the ratios of positive to negative examples between training data and real-world use of the classifier.[26]

3. Results and Discussion

3.1. *Analysis of Evaluation Methodologies*

We have evaluated the performance of two different classifiers (SVM and RF) with six different types of features using two different evaluation protocols (K-fold and LOPO cross-validation) over two different datasets (HIV and Adenovirus). The results of this combinatorial analysis are shown in the form of PR curves in Fig. 3 for the Human-HIV and the Human-Adenovirus datasets. This figure also shows the values of the mean and standard deviation of AUC-PR for different models.

With the standard K-fold cross-validation for the Human-HIV dataset we observe maximum AUC-PR scores of 58%. This score is obtained using the PSFM evolutionary features with an RF classifier (see Fig. 3(c)). However, with LOPO cross-validation, the corresponding score is 8% as shown in Fig. 3(d). A similar trend is observed for the Human-Adenovirus dataset as shown in Figs. (3(e))–(3(h)).

We also observed the same effect in ROC curves (results available on authors' supplementary information webpage: http://faculty.pieas.edu.pk/fayyaz/hpi/index.html). In order to present a clear comparison between

K-fold and LOPO cross-validation schemes, we present radar plots of the AUC-ROC over the two evaluation schemes for all models in Fig. 4. These plots clearly show that results computed with the K-fold cross-validation scheme are consistently higher than those from LOPO cross-validation for all classifiers, feature representations, datasets and performance metrics. This lends clear support to our hypothesis that overlap between training and testing data of the prediction in K-fold cross-validation leads to inflated measurement of generalization performance. These results are in line with the observations of the studies by Park and Marcotte[25] and Hamp and Rost[26] for problems involving paired data and general intra-species protein interactions, respectively. This shows that K-fold cross-validation is ineffective in controlling the redundancy between training and test data

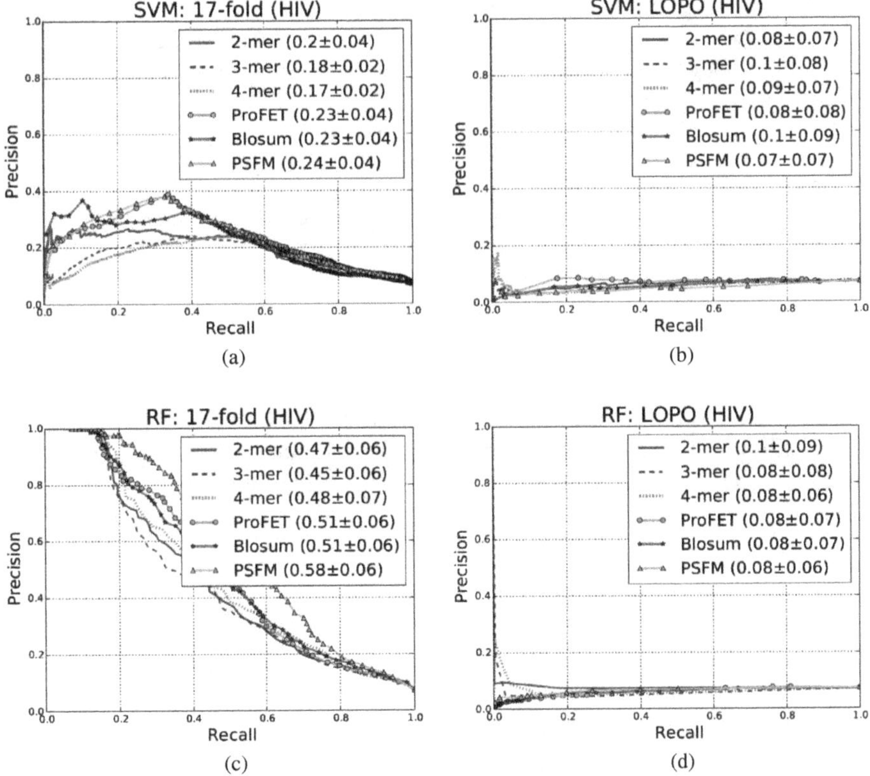

Fig. 3. Precision-recall graphs of K-fold and LOPO cross-validation for Human-HIV (a)–(d) and Human-Adenovirus (e)–(h) interaction datasets. Area under the curve is shown in parenthesis as mean $\pm$ standard deviation across different folds.

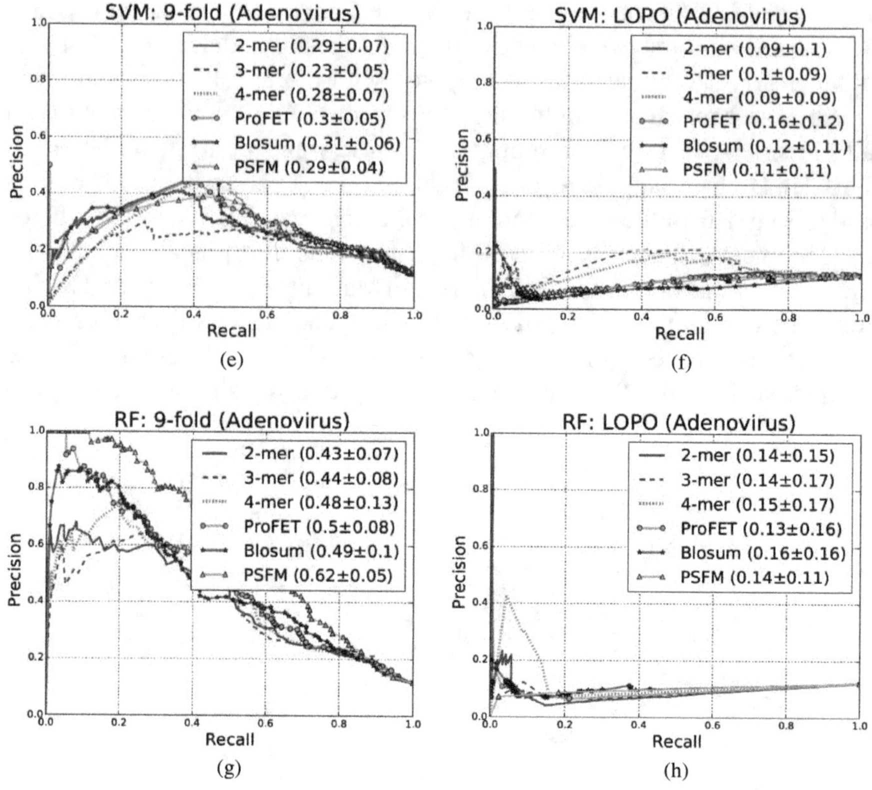

Fig. 3. (*Continued*)

for host–pathogen PPIs predictors. The underlying assumption in K-fold cross-validation is that interactions of both host and pathogen proteins in a novel test example are known during training. This assumption is restrictive for most interaction studies. As a consequence, a machine learning model optimized on the basis of K-fold cross-validation cannot be expected to perform well in predicting interactions of proteins for which no training data is available. Most of the existing studies in the literature involving host–pathogen PPIs predictions use K-fold cross-validation for performance evaluation. Therefore, results from existing techniques may be misleading and should be interpreted with a grain of salt.

With the consistency in the differences in results between *K-fold* and LOPO cross-validation observed over the wide array of models, datasets and performance metrics used in our analysis, we are confident that this phenomenon can be extrapolated to other HPI predictors.

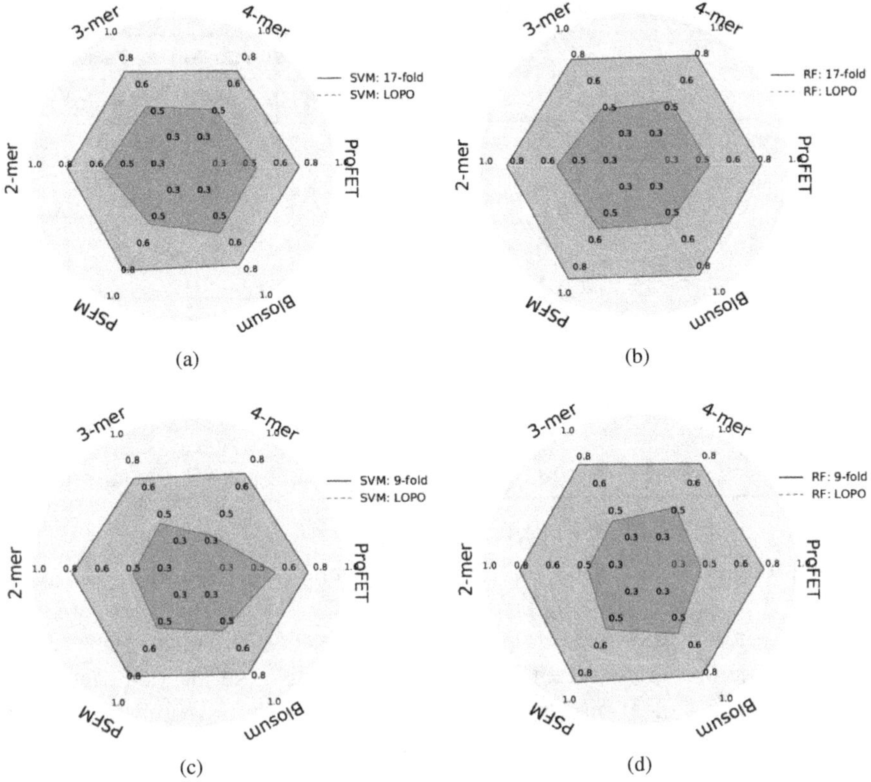

Fig. 4. Radar plots of the AUC-ROC over the two evaluation schemes for all models: (a) SVM and (b) Random Forest classifiers for the HIV dataset. Results using (c) SVM and (d) Random Forest classifiers for the Adenovirus data.

3.2. *Metrics for HPI Prediction*

We have also computed THR, FHR and MRFPP for both interaction datasets using LOPO cross-validation. These results are shown in Table 1.

For the human-HIV dataset, we observed a maximum THR of 29% with the 3-mer features using both SVM and RF classifiers. This indicates that the top scoring example for 5 out of 17 HIV proteins (i.e. ∼29%) in their corresponding folds in LOPO cross-validation is indeed a positive example. The optimum value of FHR is 1% which means that, on average, ∼1% of negative examples rank higher than the top scoring positive example for any pathogen protein. Similarly, the best MRFPP of 4 reveals that for 50% of the pathogen proteins, a positive example occurs within the

Table 1. Proposed metrics over LOPO cross-validation for all models.

	Features	Human-HIV			Human-Adenovirus		
		THR	FHR	MRFPP	THR	FHR	MRFPP
SVM	2-mer	11	04	12	00	17	34
	3-mer	29	05	15	00	19	24
	4-mer	18	04	04	00	28	44
	ProFET	06	07	19	22	07	08
	Blosum	18	06	13	00	16	13
	PSFM	06	09	22	11	20	24
RF	2-mer	24	01	04	22	20	07
	3-mer	29	02	04	22	12	04
	4-mer	24	05	09	33	12	19
	ProFET	18	10	19	11	09	11
	Blosum	18	08	19	22	08	17
	PSFM	12	05	08	22	04	04

top 4 predictions. Results for the human-adenovirus dataset in terms of these metrics are also included in Table 1.

In this study, these metrics are not used as performance measures to compare predictive power of the classifiers but to give an idea that these measures offer a clear intuition to the biologist in experimental design. We have not attempted to optimize these performance metrics.

4. Conclusions and Future Work

We investigated the effect of evaluation methodologies on the performance of the predictors built for host–pathogen PPI prediction. Through a series of simulation experiments, we showed that the results of K-fold cross-validation, which is typically employed for performance evaluation of such predictors, are overestimates of the generalization performance. This is particularly true for cases in which the user is interested in predicting interactions for proteins with no known interactions available during model construction and training.

We strongly advocate the use of leave one protein out evaluation for this purpose. We also recommend that biologist-centric performance metrics such as THR, FHR and MRFPP proposed in this work be reported for HPI predictors in conjunction with the traditional metrics like AUC-ROC and AUC-PR. In addition to the proposed cross-validation scheme and metrics, biologically meaningful analyses should also be employed to validate HPI predictions. Such analyses can include disease, gene ontology and pathway enrichment studies.[10,11,47–49] On a broader scale, this work points out a need for developing specialized machine learning methods that model the true biological nature of HPI prediction at all levels of their design with a focus on generalization performance.

Acknowledgments

Authors are thankful to Dr. Asa Ben-Hur, Colorado State University, USA for his useful suggestions and assistance in manuscript preparation. We would also like to thank Mr. Naveed Akhtar, High Performance Computing (HPC) lab, PIEAS, Pakistan for technical support and the anonymous reviewers for their valuable comments. Wajid A. Abbasi would like to acknowledge his funding from the Higher Education Commission (HEC) of Pakistan.

Supplementary Information

Details about downloading the data used in this work and supplementary information is available at: http://faculty.pieas.edu.pk/fayyaz/hpi/index. html.

References

1. National Institute of Health, Biological Sciences Curriculum Study, Understanding Emerging and Re-emerging Infectious Diseases, 1–46, 2007.
2. World Health Organization, *The Global Burden of Disease: 2004 Update*, 2008.
3. Dyer MD, Murali TM, Sobral BW, Computational prediction of host–pathogen protein–protein interactions, *Bioinformatics* **23**:i159–i166, 2007.
4. Dyer MD, Murali TM, Sobral BW, The landscape of human proteins interacting with viruses and other pathogens, *PLoS Pathog* **4**:e32, 2008.
5. Read TD *et al.*, The genome sequence of *Bacillus anthracis* Ames and comparison to closely related bacteria, *Nature* **423**:81–86, 2003.
6. Kim MS *et al.*, A draft map of the human proteome, *Nature* **509**:575–581, 2014.
7. Nourani E, Khunjush F, Durmuş S, Computational approaches for prediction of pathogen–host protein–protein interactions, *Front Microbiol* **6**:1–10, 2015.

8. Bock JR, Gough DA, Predicting protein–protein interactions from primary structure, *Bioinformatics* **17**:455–460, 2001.

9. Dyer MD, Murali TM, Sobral BW, Supervised learning and prediction of physical interactions between human and HIV proteins, *Infect Genet Evol* **11**:917–923, 2011.

10. Tastan O, Qi Y, Carbonell JG, Klein-Seetharaman J, Prediction of interactions between HIV-1 and human proteins by information integration, *Pac Symp Biocomput* 516–527, 2009.

11. Kshirsagar M, Carbonell J, Klein-Seetharaman J, Multitask learning for host–pathogen protein interactions, *Bioinformatics* **29**:i217–i226, 2013.

12. Zhang SW, Wei ZG, Some remarks on prediction of protein–protein interaction with machine learning, *Med Chem* **11**:254–264, 2015.

13. Arnold R, Boonen K, Sun MGF, Kim PM, Computational analysis of interactomes: Current and future perspectives for bioinformatics approaches to model the host–pathogen interaction space, *Methods* **57**:508–518, 2012.

14. Ben-Hur A, Noble WS, Choosing negative examples for the prediction of protein–protein interactions, *BMC Bioinformatics* **7**:1–6, 2006.

15. Shen J *et al.*, Predicting protein–protein interactions based only on sequences information, *Proc Natl Acad Sci* **104**:4337–4341, 2007.

16. Cui G, Fang C, Han K, Prediction of protein–protein interactions between viruses and human by an SVM model, *BMC Bioinformatics* **13**:10, 2012.

17. Hamp T, Rost B, Evolutionary profiles improve protein–protein interaction prediction from sequence, *Bioinformatics* **31**:1945–1950, 2015.

18. Zahiri J, Yaghoubi O, Mohammad-Noori M, Ebrahimpour R, Masoudi-Nejad A, PPIevo: Protein–protein interaction prediction from PSSM based evolutionary information, *Genomics* **102**:237–242, 2013.

19. Barman RK, Saha S, Das S, Prediction of interactions between viral and host proteins using supervised machine learning methods, *PLoS ONE* **9**:e112034, 2014.

20. Qi Y, Tastan O, Carbonell JG, Klein-Seetharaman J, Weston J, Semi-supervised multi-task learning for predicting interactions between HIV-1 and human proteins, *Bioinformatics* **26**:i645–i652, 2010.

21. Zhou H, Jin J, Wong L, Progress in computational studies of host–pathogen interactions, *J Bioinform Comput Biol* **11**:1–26, 2013.

22. Emamjomeh A, Goliaei B, Zahiri J, Ebrahimpour R, Predicting of protein–protein interactions between human and hepatitis C virus via an ensemble learning method, *Mol Biosyst* **10**:3147–3154, 2014.

23. Park Y, Critical assessment of sequence-based protein–protein interaction prediction methods that do not require homologous protein sequences, *BMC Bioinformatics* **10**:419, 2009.

24. Wagstaff KL, Machine learning that matters, *29th International Conference on Machine Learning*, California Institute of Technology, pp. 529–536, 2012.

25. Park Y, Marcotte EM, A flaw in the typical evaluation scheme for pair-input computational predictions, *Nat Methods* **9**:1134–1136, 2012.

26. Hamp T, Rost B, More challenges for machine learning protein interactions, *Bioinformatics* **31**:1521–1525, 2015.

27. Alpaydin E, Design and analysis of machine learning experiments, in *Introduction to Machine Learning*, 2nd ed., Cambridge, MA, MIT Press, pp. 486–488, 2010.

28. Davis J, Goadrich M, The Relationship between Precision-Recall and ROC Curves, *Proc 23rd Int Conf Machine Learning*, ACM, pp. 233–240, 2006.

29. Minhas F, Ben-Hur A, Multiple instance learning of Calmodulin binding sites, *Bioinformatics* **28**:i416–i422, 2012.

30. Fu W *et al.*, Human immunodeficiency virus type 1, human protein interaction database at NCBI, *Nucleic Acids Res* **37**:D417–422, 2009.

31. Durmuş TS *et al.*, PHISTO: Pathogen–host interaction search tool, *Bioinformatics* **29**:1357–1358, 2013.

32. Huang Y, Niu B, Gao Y, Fu L, Li W, CD-HIT Suite: A web server for clustering and comparing biological sequences, *Bioinformatics* **26**:680–682, 2010.

33. Cortes C, Vapnik V, Support-vector networks, *Mach Learn* **20**:273–297, 1995.

34. Breiman L, Random forests, *Mach Learn* **45**:5–32, 2001.

35. Minhas F, Geiss BJ, Ben-Hur A, PAIRpred: Partner-specific prediction of interacting residues from sequence and structure, *Protein Struct Funct Bioinf* **82**:1142–1155, 2014.

36. Ben-Hur A, Ong CS, Sonnenburg S, Schölkopf B, Rätsch G, Support vector machines and kernels for computational biology, *PLoS Comput Biol* **4**: e1000173, 2008.

37. Wuchty S, Computational prediction of host-parasite protein interactions between, *P. Falciparum and H. Sapiens PLoS ONE* **6**:e26960, 2011.

38. Pedregosa F *et al.*, Scikit-learn: Machine learning in python, *J Mach Learn Res* **12**:2825–2830, 2011.

39. Leslie C, Eskin E, Noble WS, The spectrum kernel: A string kernel for SVM protein classification, *Pac Symp Biocomput*, pp. 564–575, 2002.

40. Ofer D, Linial M, ProFET: Feature engineering captures high-level protein functions, *Bioinformatics* **31**:3429–3436, 2015.

41. Eddy SR, Where did the BLOSUM62 alignment score matrix come from? *Nat Biotech* **22**:1035–1036, 2004.

42. Aumentado-Armstrong TT, Istrate B, Murgita RA, Algorithmic approaches to protein–protein interaction site prediction, *Algorithms Mol Biol* **10**:1–21, 2015.

43. Zaki N, Lazarova-Molnar S, El-Hajj W, Campbell P, Protein–protein interaction based on pairwise similarity, *BMC Bioinformatics* **10**:150, 2009.

44. Westen GJV *et al.*, Benchmarking of protein descriptor sets in proteochemometric modeling (part 1): Comparative study of 13 amino acid descriptor sets, *J Cheminform* **5**:41, 2013.

45. Altschul SF *et al.*, Gapped BLAST and PSI-BLAST: A new generation of protein database search programs, *Nucleic Acids Res* **25**:3389–3402, 1997.

46. Pruitt KD, Tatusova T, Maglott DR, NCBI Reference Sequence (RefSeq): A curated non-redundant sequence database of genomes, transcripts and proteins, *Nucleic Acids Res* **33**:D501–D504, 2005.

47. Zhou H *et al.*, Stringent homology-based prediction of *H. Sapiens–M. Tuberculosis* H37Rv protein–protein interactions, *Biol Direct* **9**:1–30, 2014.

48. Rapanoel HA, Mazandu GK, Mulder NJ, Predicting and analyzing interactions between *Mycobacterium tuberculosis* and its human host, *PLoS ONE* **8**:e67472, 2013.

49. Durmuş TS, Çakir T, Ülgen KÖ, Infection strategies of bacterial and viral pathogens through pathogen–human protein–protein interactions, *Front Microbiol* **3**:1–11, 2012.

Wajid Arshad Abbasi is currrntly pursuing his PhD in the Department of Computer and Information Sciences, Pakistan Institute of Engineering and Applied Sciences (PIEAS), Islamabad, Pakistan. He is a recipient of the indigenous PhD fellowship by the Higher Education Commission (HEC) of Pakistan. He received his M.Phil degree in Software Engineering from the Iqra University, Karachi, Pakistan, in 2013. He also holds a Masters in Computer Science from the University of Azad Jammu and Kashmir, Muzaffarabad, Pakistan.

Dr. Fayyaz Ul Amir Afsar Minhas is a senior scientist with the Department of Computer and Information Sciences, Pakistan Institute of Engineering and Applied Sciences (PIEAS), Islamabad, Pakistan. Dr. Minhas received his PhD degree in Bioinformatics from Colorado State University, USA on a Fulbright Scholarship. He has also been awarded the National Youth Award by the Government of Pakistan for his contributions in science and technology. His lab focuses on applications of machine learning in Bioinformatics and the analysis of biomedical data. His webpage is available at the URL: http://faculty.pieas.edu.pk/fayyaz/.

Part III
Protein Function Prediction from Protein-Protein Interactions

CHAPTER 9

Protein Function Prediction from Protein–Protein Interaction Network Using Gene Ontology Based Neighborhood Analysis and Physico-Chemical Features[a]

Sovan Saha[*,§], Abhimanyu Prasad[*,¶], Piyali Chatterjee[†,‖],
Subhadip Basu[‡,**] and Mita Nasipuri[‡,††]

*Department of Computer Science & Engineering
Dr. Sudhir Chandra Sur Degree Engineering College
540 Dum Dum Road, Near Dum Dum Jn. Station, Surermath
Kolkata 700074, India

†Department of Computer Science & Engineering
Netaji Subhash Engineering College
Techno City Panchpota, Garia, Kolkata 700152, India

‡Department of Computer Science & Engineering
Jadavpur University, 188, Raja S.C. Mallick Road
Kolkata 700032, India
§sovansaha12@gmail.com
¶prasad.abhi150@gmail.com
‖chatterjee_piyali@yahoo.com
**bsubhadip@gmail.com
††mitanasipuri@yahoo.com

Protein Function Prediction from Protein–Protein Interaction Network (PPIN) and physico-chemical features using the Gene Ontology (GO) classification are indeed very useful for assigning biological or

---

[a]This article was previously published in *Journal of Bioinformatics and Computational Biology*. Vol: 16, No. 6 (2018), 1850025 (15 pages).
§Corresponding author.

biochemical functions to a protein. They also lead to the identification of those significant proteins which are responsible for the generation of various diseases whose drugs are still yet to be discovered. So, the prediction of GO functional terms from PPIN and sequence is an important field of study. In this work, we have proposed a methodology, **Multi Label Protein Function Prediction (ML_PFP)** which is based on Neighborhood analysis empowered with physico-chemical features of constituent amino acids to predict the functional group of unannotated protein. A protein does not perform functions in isolation rather it performs functions in a group by interacting with others. So a protein is involved in many functions or, in other words, may be associated with multiple functional groups or labels or GO terms. Though functional group of other known interacting partner protein and its physico-chemical features provide useful information, assignment of multiple labels to unannotated protein is a very challenging task. Here, we have taken *Homo sapiens* or Human PPIN as well as *Saccharomyces cerevisiae* or yeast PPIN along with their GO terms to predict functional groups or GO terms of unannotated proteins. This work has become very challenging as both Human and Yeast protein dataset are voluminous and complex in nature and multi-label functional groups assignment has also added a new dimension to this challenge. Our algorithm has been observed to achieve a better performance in Cellular Function, Molecular Function and Biological Process of both yeast and human network when compared with the other existing state-of-the-art methodologies which will be discussed in detail in the results section.

Keywords: Protein–protein interaction network; protein cluster; physico-chemical features; gene ontology; biological process; cellular function; molecular function.

1. Introduction

With the advent of high-throughput human genome projects, accumulation of increasing amount of biological data necessitates the development of computational techniques for assigning functional group to proteins in faster and cost-effective manner because wet lab techniques are accurate but costly and have low throughput. This intelligent hypothetical functional annotation scheme not only provides clues about protein functions but also is helpful in understanding sequence similarity, Pathway analysis, Phylogenetic analysis, Functional interactions, etc.[1,2]

As a protein is likely to be involved in a number of functional groups, so protein function prediction problem can be viewed as a multilabel classification problem. Thus this problem becomes a more challenging task when the number of functional groups or labels is large in numbers and unbalanced, too. Moreover, functional groups are not independent but structured in a hierarchy and in some cases, they are incomplete or missing for some proteins making the overall problem difficult and challenging.

Several computational techniques have been developed from the protein sequence,[3,4] protein structure,[5,6] Protein–Protein Interaction Network (PPIN) data,[7–10] motif or signature.[11,12]

Vast amounts of PPIN data produced by high-throughput techniques can also be a part of Protein Functions as a number of proteins are involved in a particular functional group. Predictions based on PPIN, represented as graph (where every node corresponds to an individual protein and each edge between a pair of nodes corresponds to interaction between them) assign functions to a protein with an assumption that closer neighbors of a protein are functionally similar.

In the literature review, according to Sharan *et al.*,[13] function prediction from PPIN can be classified into two approaches: direct annotation and module-assisted schemes. In direct annotation scheme, protein functions are inferred from its connections. Neighborhood counting, Graph-theoretic methods, Markov Random fields and integrating multiple information sources fall under this category. Neighborhood counting method determines functions based on its level-1 neighbors. Schwikowski,[9] proposed a simple and effective neighborhood-counting method to allocate functions to a protein based on the number of occurrences of functional labels in protein interaction neighborhood. While in the chi-square method,[14] functions are assigned to a protein based on the largest chi-square scores. For a protein P, each function "f" is assigned a score

$$\frac{(n_f - e_f)^2}{e_f},\tag{1}$$

where n_f is the number of proteins in the n-neighborhood of P that have the function f and e_f is the expectation of this number based on the frequency of f among all proteins in the network. Graph-theoretic methods view this problem as graph analysis by applying different algorithms like cut-based and flow-based algorithms. Many graph related algorithms, e.g. graph connectivity, flow-based approach, etc., are also seen to be applied for the analysis of function prediction in the work of Vazquez *et al.*[8] and

Nabieva *et al.*[15] The assumption that the function of a protein is independent of all other proteins given the functions of its immediate neighbors leads researchers to apply Markov random field. Letovsky and Kasif[16] incorporate the assumption of a binomial model for neighbors in locality of a protein annotated with respect to a given time period. Similarly, Wu comes in with a new probabilistic model to predict unannotated function of proteins from the structure of the PPIN.[17] On the other hand, module-assisted schemes identify modules of related proteins and then annotate each module based on the known functions of its members. Module assisted techniques are different. Once the module is detected, function of the module is easily predicted. So, module finding algorithms can be divided into methods using solely network topology information and methods that utilize addition data sources, such as gene expression data. Bader and Hogue[18] propose to identify denser regions by some heuristic parameters through molecular complex detection (MCODE). For detection of modules, Hierarchical clustering and Graph clustering methods can be mentioned. Selection of similarity measure between protein pairs is the key point in hierarchical methods. In the work of Samanta *et al.*,[19] more significance is given to the number of common interacting partners between two connected proteins, based on the assumption that if they have connections then they share a common functionality.

Chen *et al.* take this neighborhood property to next hierarchical level by bringing in the concept of functional similarity between a protein and its direct as well as indirect neighbors.[11] Different types of graph-clustering algorithms have been applied to the PPIN. Spirin and Mirny[20] execute the concept of super paramagnetic clustering and a Monte Carlo algorithm in this field of function prediction.

A significant proportion of PPIN obtained from high-throughput experiments have been found to contain false positive and false negative data which suffer from low recall and high precision values in most of the cases. This has led some researchers to use integrated data from heterogeneous sources like network topology, sequence, gene expression, domain information, protein complexes and others to predict protein function from PPIN. Xiong *et al.*[10] combines PPIN information and protein sequence information to increase the performance of the predictor. They add implicit edges to the network with explicit or existing edges and employ a collective classification algorithm to predict the function. In our previous work,[21–23] neighborhood properties and physico-chemical features collectively determine the functional group.

Of late, during the last couple of years, protein function prediction problem has been characterized by several factors like diversity in functional groups, hierarchical relationships among functional groups and incomplete or missing information about them. Thus it is viewed as multi-label learning algorithms.[24–26] Hierarchical relationships among labels are found in MIPS Functional Catalogue and Gene Ontology. Valentini[25] uses a binary classifier for each label according to True Path Rule (TPR) and the FunCat. In recent work of Guoxian and the co-authors,[27] they explored the incomplete label problem in a hierarchical manner using function correlation. Piovesan *et al.*[28] propose a unique approach of combining PPI information, domain information and sequence information for the prediction of protein function. Zhao *et al.*[2] creates a dynamic weighted interactome network enriched with PPIN, time course gene expression data, protein's domain information and protein complex information from which prediction is executed using a ranking methodology.

All these methods discussed above have already left a mark in this field of protein function prediction which can be further used in the identification of essential drug targets (essential gene or protein identification).[29] Still detailed study and analysis have revealed the fact that there is lot of scope for improvement in this field related to PPIN. PPIN generated by high-throughput techniques may contain false positive data resulting in unsatisfactory recall and precision scores. This motivates us to reconstruct the PPIN by removing less significant edges. The fact that the multi-label classification can be improved by employing both significant PPIN and protein physico-chemical features actually motivates us to design an algorithm **ML_PFP**. But this method ignores the incompleteness or missing values in annotations in order to avoid computational complexity as human and yeast data are voluminous.

Proposed work can be described in two phases: (1) Each protein in human and yeast PPIN is considered to be unannotated and is added to the test set. (2) In the second phase, test set data (proteins) are predicted using neighborhood analysis of clusters and protein sequence information, i.e. physico-chemical features. The two phases are completely unique in terms of their approach. This process continues until and unless all proteins in both the networks of yeast and humans get annotated.

The overall pipeline of the proposed **ML_PFP** is mainly dependent on PPIN and protein sequences. These two major components are biologically very relevant in this area of research. The same has been also observed in the other related successful works which are specifically based on these two. Some of the important ones are highlighted in Table 1. It should be noted that in Table 1, these two are not only restricted to protein function prediction but also help in essential disease gene identification which ultimately leave a greater impact in this biological

Table 1. Computational studies based on PPIN and protein sequences.

Utilized feature	Description	References
Protein sequence	Prediction of changes in amino acid sequences affecting protein function prediction.	3
	Detection of protein functions and interactions from genomic sequences.	4
PPIN	Protein function prediction using neighbor relativity.	7
	Function prediction from PPIN can be classified into two approaches: direct annotation scheme and module-assisted schemes.	13
	Assignment of proteins to functional classes on the basis of their network of physical interactions as determined by minimizing the number of protein interactions among different functional categories.	8
	Protein function prediction from protein–protein interaction by the application of Apriori algorithm.	32
	Assignment of multiple GO terms (protein functions) to the unannotated proteins using ranked methodology based on protein–protein interaction data.	39
	Two new methods are proposed. FunPred 1.1 uses a combination of three simple-yet-effective neighborhood scoring techniques. FunPred 1.2 applies a heuristic approach using the edge clustering coefficient to reduce the search space.	21
PPIN, protein domain and sequence	Combination of the information obtained from protein–protein interaction, domain and sequence for the prediction of protein function.	28

Table 1. (*Continued*)

Utilized feature	Description	References
Physico-chemical features of amino acid sequence of proteins	Disease gene identification based on physico-chemical features of amino acids.	40
	Functional prediction of amino acid substitutions based on physico-chemical properties.	41
PPIN and Physico-chemical features of amino acid sequence of proteins	Protein Function Prediction from PPIN using Physico-chemical features of Amino Acids.	23
	A classifier has been developed which uses decision tree for classification process. The protein function is predicted on the basis of matched physico-chemical features per each protein function.	34

field of study. So these properties have been selected to develop the basic outlay of the proposed model **ML_PFP**.

Before proceeding into the main section of the work, various terminologies like PPIN,[21,22,30–32] Protein Cluster, Sub-graph, Level-1 Neighbors,[21] Edge Weight,[21,33] Node Weight,[21,33] Physico-chemical Features,[23,34] Participation Score, XGBoost Classifier,[35] Random forest Classifier,[36] Extra Tree Classifier[37,38] and Recursive feature elimination (RFE) Classifier[38] need to be understood, which are described in detail in the supplementary document.

2. Dataset

Gene Ontology (GO) dataset of humans and yeast has been considered in**ML_PFP**. All these data are collected from UniProt.[42] Three categories: Cellular-component, Molecular-function and Biological-process are engaged in the GO system. In this system, each protein may be annotated by several GO terms (like GO: 0000016) in each category. Table 2 highlights the detailed analysis of PPIN of human and yeast which reveals the fact that PPIN of humans is relatively denser in comparison to PPIN of yeast. But, it has been observed that interactions of most of the proteins in the PPIN of human (see Table S3 in supplementary document) are still

Table 2. Statistics of PPIN of *Saccharomyces cerevisiae* and *Homo sapiens*.

Organism	Number of proteins	Number of interactions	GO terms	Cellular component	Molecular function	Biological process
Yeast	2130	5282	4352	883	1072	2397
Human	11946	54283	17009	1823	3721	11465

unknown in UniProt database in comparison to yeast (see Table S4 in supplementary document) due to which precision, recall and F-score of all methodologies including ours (see Sec. 4) falls for PPIN of humans when compared to the same for the PPIN of yeast.

3. Proposed Method

In our method **ML_PFP**, each protein in the network of yeast and humans is considered to be unannotated and is added to the unknown protein set P_U. For a particular protein in P_U, it selects refined level-1 neighbors after application of node and edge weight. Corresponding protein cluster for every refined level-1 neighbor of that protein is formed which comprises of their immediate neighbors provided that incorporating such edges in the cluster[33] does not cause the edge weight to drop below the assigned threshold value α. Thus, for a target protein having say, k numbers of refined level-1 neighbors, equivalent k numbers of protein clusters are generated. Next the summation of participation score ($\sum \text{Par\_Score}_{P_j}^{C_{i}}$) of all proteins in each cluster is calculated. The unannotated protein is assigned to the cluster having highest participation score. Consequently, unannotated proteins get associated with multiple functional groups of all proteins in that cluster. In order to find out significant functional groups, those that may be labeled on unannotated proteins, **ML_PFP** exploits the physico-chemical properties of protein of interest and all the member proteins in that cluster.

To do that, **ML_PFP** computes $\text{PCP}_{\text{score}}$ of unannotated protein and all proteins of the chosen cluster. Finally, multiple GO terms of member protein are assigned to unannotated protein considering the fact that minimum difference in $\text{PCP}_{\text{score}}$ is found between the unannotated and member protein. The entire methodology of **ML_PFP** is outlined in Algorithm 1 as well as pictorially represented in Figs. 1 and 2.

Algorithm 1. Methodology of ML_PFP

Input: Dataset containing protein pair with multi-functional label (GO).
          Set threshold α and β for minimum edge weight and node
          weight respectively.
Output: Functional group of un-annotated proteins

---

  begin
  //unannotated protein selection forming set P_U
  for each network of human and yeast
      Consider each protein to be unannotated and add it to set P_U.
    //network refinement
    for every edges between u and v in G
        Compute edge weight
        if edge weight $W_{uv} = 0$, then remove corresponding edge.
    for every nodes in G
        Compute node weight
        if node weight does not exceed threshold β remove corresponding node
    //end of network refinement
    //unannotated protein function prediction
    for all proteins in set P_U
    //construct clusters of every level-1 *neighbors*
    for each level-1 neighbor K
        Make it a seed of cluster C_i i.e. $C_i = \{K\}$
        Add neighbors of K to C_i such that inclusion of neighbors of K
        does not cause edge weight to fall below α i.e. $C_i = \{K\} \cup N_K$;
        $G = G - C_i$.
    //end of formation of cluster
     //calculation of participation score
    for all protein in each cluster
        Compute $\text{Par\_Score}_{P_j}^{C_i,}$
        Obtain $\sum \text{Par\_Score}_{P_j}^{C_i,}$
    //end of calculation of participation score
    //Assignment of cluster to unannotated protein along with all GO
    Assign the unannotated protein to the cluster C_i having highest $\sum \text{Par}_{\text{Score}}^{C_i, P_j}$
    //Assignment of essential GO to unannotated protein
    for each protein P_j in the assigned clusters C_i
        Find the difference of $\text{PCP}_{\text{score}}$ of unannotated protein and $\text{PCP}_{\text{score}}$
        of the rest of the proteins in C_i.
    Assign the unannotated protein to the associated GO's of P_j
    which has minimum difference in $\text{PCP}_{\text{score}}$.
    //end of unannotated protein function prediction
    Consider the next protein to be unannotated and add it to set P_U.
    Repeat all the above steps until and unless all the proteins in the network
    of humans and yeast get annotated.
    //end of unannotated protein selection
 end

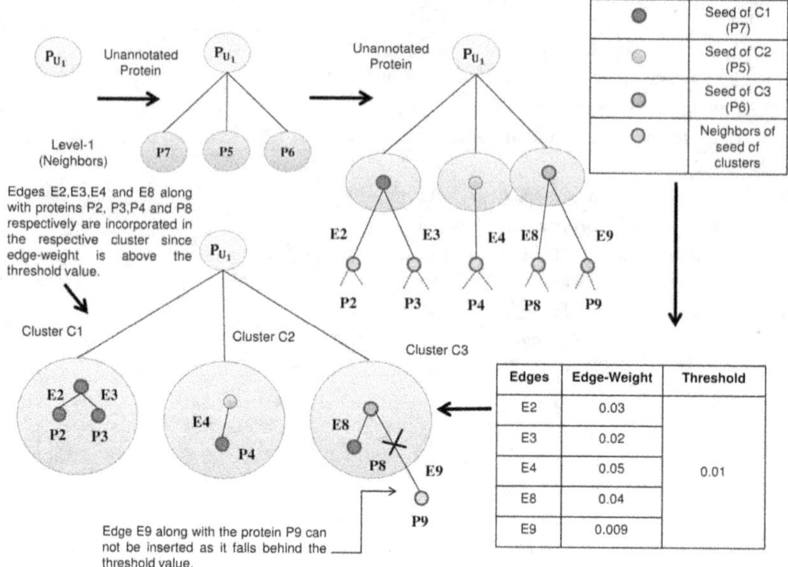

Fig. 1. Schematic diagram of formations of clusters in ML_PFP.

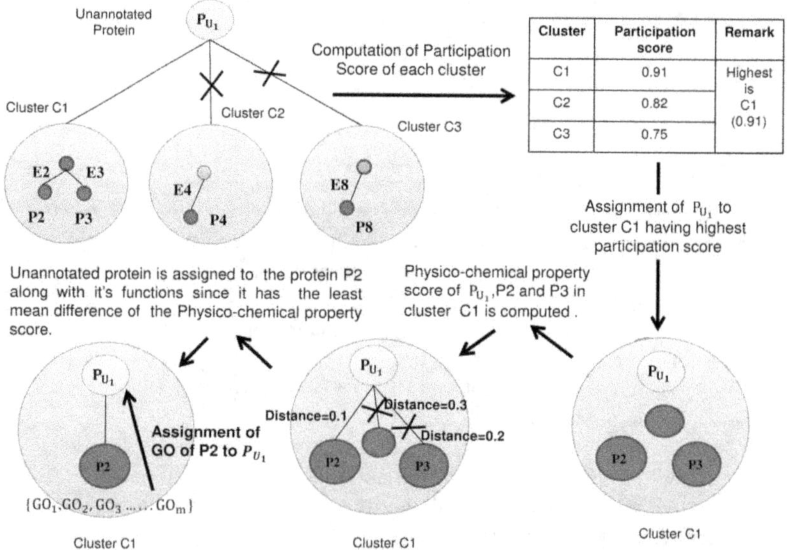

Fig. 2. Schematic diagram of unannotated protein function prediction in ML_PFP.

4. Results and Discussion

Result analysis of **ML_PFP** includes all the proteins as test samples in the entire human and yeast GO dataset. After successful prediction of our test samples by our algorithm, its corresponding result is analyzed using Precision (P_i), Recall (R_i) and F-Score (F_i).[21] The overall performance evaluation has been highlighted in Table 3 for all the three functional categories of human and yeast GO dataset.

Proper selection and variation of threshold value for node weight (β) and edge weight (α) also plays a key role in the generation of results as displayed in Table 3. In this proposed work, three different thresholds have been considered: High, Medium and Low.[43] Best result has been generated for high threshold since it ensures the presence of maximum number of essential proteins (by the application of node and edge weight) while a fall in the accuracy level has been observed both for medium and low threshold due to presence of noisy (unreliable proteins) proteins. An overall display of the entire scenario has been represented in Tables S1 and S2 (see supplementary document(online)) for human and yeast, respectively.

Besides not all the physico-chemical features have been considered for protein function prediction. Six different top-ranked physico-chemical features: aliphatic index, gravy, aromacity, number of negatively charged residues, numbers of positively charged residues and isoelectric point are considered for computing $\mathrm{PCP_{score}}$. These top-ranked features are selected from 10 different physico-chemical features (see supplementary document) by the execution of four classifiers: XGBoost classifier, Random Forest classifier, Extra Tree classifier and RFE classifier. At first, top five out of 10 ranked features have been selected by each classifier. Then from these selected features, maximum number of hits for each feature has been noted from which ultimate top six ranked features get chosen.

Table 3. Result analysis, precision, Recall and F-Score obtained in our methodology ML_PFP in the PPIN of both human and yeast.

Organism	Functional category	Precision	Recall	F-score
Yeast	Cellular component	0.81	0.64	0.72
	Molecular function	0.69	0.51	0.59
	Biological process	0.75	0.61	0.67
Human	Cellular component	0.60	0.45	0.51
	Molecular function	0.29	0.26	0.27
	Biological process	0.35	0.24	0.28

Figures S2–S7 (see supplementary document) also includes other existing methodologies for unannotated protein function prediction like NRC,[7] FS-weight #1 (#1 represents only level-1 proteins),[44] FS-weight #1 (#1 and #2 represents level-1 and level-2 proteins),[44] Neighborhood counting #1 (#1 represents only level-1 proteins),[9] Neighborhood counting #1 (#1 and #2 represents level-1 and level-2 proteins),[9] Chi square #1 (#1 represents only level-1 proteins),[14] Chi square #1 (#1 and #2 represents level-1 and level-2 proteins),[14] FunPred,[21] INGA,[28] FFPred3[45] and DeepGO.[46] These are the well-known and standard methods (implemented on the same dataset of humans and yeast as of our methodology ML_PFP) which have been considered in our work for comparison of our method's performance. From Figs. S2–S7 (see supplementary document), it is observed that ML_PFP performs better when compared to existing predictors, as follows:

(1) In our work, function prediction is based on the PPIN formed by levels-1 and 2 neighbors of unannotated proteins. Now this network may contain essential as well as unreliable proteins which have been taken into consideration in our work through the application of node and edge weight. Significance of a particular neighbor protein is based on its amount of contribution in function prediction. So Participation Score has been used. All these facts have been carefully observed and implemented successfully.

(2) Our protein function prediction model is not only dependent on network-based properties but it also uses physico-chemical features of amino acid sequence of each protein. It should be noted that feature selection has also been performed which filters out the unreliable ones. These properties are vital to protein function prediction since proteins are formed from amino acid and any sort of information derived from it will definitely play a significant role to decide which functions or GO terms of which proteins will be annotated to the unknown ones.

The other existing predictors shown in Figs. S2–S7 (see supplementary document) lack these two key features of **ML_PFP** which makes it perform better among all. The work of Moosavi *et al.*,[21] clearly outperforms the rest except **ML_PFP**. For chi-square methods (Chi-square #1 and Chi-square #1 and #2), the weak prediction outcomes are generated from the sparseness of the entire PPIN. The chi-square methods work in a better way on the denser parts of the interaction network as claimed by Hishigaki *et al.*[2] Simultaneous use of both levels-1 and 2 neighbors increases the prediction performance for all other methods considered here except the Chi-square method. Though the neighborhood counting method is simple in nature, yet it succeeds in generating a notable performance, when it uses both levels-1 and 2 neighbors. None of these methods focuses on refining the PPIN by the elimination of non-essential edges. This will assure that function prediction should be always generated from more significant nodes rather than the non-significant ones. It will eventually enhance prediction accuracy as implemented in FunPred. FunPred 1.1 uses the combination of three simple-yet-effective scoring techniques: Neighborhood Score, Relative functional similarity score and Protein path connectivity score to predict unannotated protein functions. This unique combination makes it relatively a better performer than the others. While FunPred 1.2 implements the same methodology as that of FunPred 1.1, it applies a heuristic approach using the edge clustering coefficient to reduce the search space by identifying densely connected neighborhood regions. Due to this, FunPred 1.2 not only performs better than FunPred 1.1 but also than the other methods.

It is clearly visible from the above discussion that PPIN information alone is not sufficient to predict unannotated protein function. So more diversified information needs to be incorporated to improve PPIN-based prediction accuracy level. This actually motivated us to develop **ML_PFP**. We have considered another PPIN, sequence and domain-based information-based composite prediction methodology, INGA to compare with our composite sequence and PPIN-based method, **ML_PFP**. Though INGA manages to surpass Nrc method, it still remains a little behind **ML_PFP**. This is because of the presence of unreliable or non-essential proteins that always hamper the prediction which is surpassed in our method in our network refinement phase through node and edge weight.

Similar instances have also been observed in the case of FFPred3 and DeepGO. Though **ML_PFP** performs better than FFPred3 in all the sections of both human and yeast PPIN, it fails in the biological process of yeast where FFPred3 is the clear winner. Despite performing well in other datasets, DeepGO fails relatively by a huge margin when compared to **ML_PFP** and other methodologies. Moreover DeepGO fails to predict the functions of those proteins whose sequence length is greater than 1002, which is another disadvantage of it.

ML_PFP not only succeeds in predicting known protein functions but it can also predict the unknown functions of protein pairs in UniProt.

Functions of each of the two proteins in an unknown pair are predicted at first separately by **ML_PFP**. Then common functions in the predicted functions of both the proteins are assigned to their corresponding unknown pair. In spite of such ability, it fails to predict some of the unknown pairs because less or practically no interactions are available in the PPIN for that specific pair. The detailed numerical analysis of the prediction of unknown protein pairs in UniProt for both human and yeast dataset are available in Tables S3 and S4 (see supplementary document).

So, in a nutshell, this prediction methodology **ML_PFP** definitely performs well in all aspects in sparse as well as denser network of yeast and human PPIN, which in turn really proved to be beneficial in this field of study of protein function prediction. But, we have still not yet succeeded in comparing the performance of our methodology, **ML_PFP**, with that of function prediction by dynamic network in the work of Zhao *et al.*[2] So incorporation of dynamic network associated with domain, structure[47–49] as well as functional categorization might be proved as a source of relevant information which might be embedded in the prediction model in future. Irrespective of that, this prediction methodology can be applied to disease specific datasets to identify the functions of those proteins which are responsible for the cause of these diseases.[29] The supplementary document is available online at https://goo.gl/cfVfXc.

Acknowledgments

Authors are thankful to the CMATER research laboratory of the Computer Science Department, Jadavpur University, India, for providing infrastructure facilities during progress of the work. This project is partially supported by the CMATER research laboratory of the Computer Science and Engineering Department, Jadavpur University, India, and UGC Research Award (F.30-31/2016(SA-II)) from UGC, Government of India, and DBT project (No.BT/PR16356/BID/7/596/2016), Ministry of Science and Technology, Government of India.

References

1. Pesquita C, Faria D, Falcão AO, Lord P, Couto FM, Semantic similarity in biomedical ontologies, *PLoS Comput Biol* **5**:e1000443, 2009.

2. Zhao B, Wang J, Li M, Li X, Li Y, Wu FX, Pan Y, A new method for predicting protein functions from dynamic weighted interactome networks, *IEEE Trans Nanobiosci* **15**:131–139, 2016.

3. Ng PC, Henikoff S, SIFT: Predicting amino acid changes that affect protein function, *Nucleic Acids Res* **31**:3812–3814, 2003.

4. Marcotte EM, Pellegrini M, Ng HL, Rice DW, Yeates TO, Eisenberg D, Detecting protein function and protein-protein interactions from genome sequences, *Science (New York, N.Y.)* **285**:751–753, 1999.

5. Lee D, Redfern O, Orengo C, Predicting protein function from sequence and structure, *Nat Rev Mol Cell Biol* **8**:995–1005, 2007.

6. Mills, CL, Beuning PJ, Ondrechen MJ, Biochemical functional predictions for protein structures of unknown or uncertain function, *Comput Struct Biotechnol J* **13**:182–191, 2015.

7. Moosavi S, Rahgozar M, Rahimi A, Protein function prediction using neighbor relativity in protein-protein interaction network, *Comput Biol Chem* **43**:11–16, 2013.

8. Vazquez A, Flammini A, Maritan A, Vespignani A, Global protein function prediction from protein-protein interaction networks, *Nat Biotechnol* **21**:697–700, 2003.

9. Schwikowski B, Uetz P, Fields S, A network of protein-protein interactions in yeast, *Nat Biotechnol* **18**:1257–1261, 2000.

10. Xiong W, Liu H, Guan J, Zhou S, Protein function prediction by collective classification with explicit and implicit edges in protein-protein interaction networks, *BMC Bioinform* **14**:S4, 2013.

11. Chen J, Hsu W, Lee ML, Ng S-K, Labeling network motifs in protein interactomes for protein function prediction, *2007 IEEE 23rd Int Conf Data Eng*, pp. 546–555 (IEEE, 2007), doi: 10.1109/ICDE.2007.367900.

12. Lichtarge O, Bourne HR, Cohen FE, An evolutionary trace method defines binding surfaces common to protein families, *J Mol Biol* **257**:342–358, 1996.

13. Sharan R, Ulitsky I, Shamir R, Network-based prediction of protein function, *Mol Syst Biol* **3**:88, 2007.

14. Hishigaki H, Nakai K, Ono T, Tanigami A, Takagi T, Assessment of prediction accuracy of protein function from protein–protein interaction data, *Yeast (Chichester, England)* **18**:523–531, 2001.

15. Nabieva E, Jim K, Agarwal A, Chazelle B, Singh M, Whole-proteome prediction of protein function via graph-theoretic analysis of interaction maps, *Bioinformatics (Oxford, England)* **21** Suppl 1:i302–i310, 2005.

16. Letovsky S, Kasif S, Predicting protein function from protein/protein interaction data: A probabilistic approach, *Bioinformatics* **19**:i197–i204, 2003.

17. Wu DD, An efficient approach to detect a protein community from a seed, *2005 IEEE Symp Comput Intell Bioinform Comput Biol*, pp. 1–7 (IEEE, 2005), doi: 10.1109/CIBCB.2005.1594909.

18. Bader GD, Hogue CWV, An automated method for finding molecular complexes in large protein interaction networks, *BMC Bioinform* **27**:1–27, 2003.

19. Samanta MP, Liang S, Predicting protein functions from redundancies in large-scale protein interaction networks, *Proc Natl Acad Sci USA* **100**:12579–12583, 2003.

20. Spirin V, Mirny LA, Protein complexes and functional modules in molecular networks, *Proc Natl Acad Sci USA* **100**:12123–12128, 2003.

21. Saha S, Chatterjee P, Basu S, Kundu M, Nasipuri M, FunPred-1: Protein function prediction from a protein interaction network using neighborhood analysis, *Cellular and Mol Biol Lett* **19**:675–691, 2014.

22. Saha S, Chatterjee P, Basu S, Kundu M, Nasipuri M, Improving prediction of protein function from protein interaction network using intelligent neighborhood approach, *2012 Int Conf Commun Devices Intell Syst*, pp. 584–587 (IEEE, 2012), doi: 10.1109/CODIS.2012.6422270.

23. Saha S, Chatterjee P, Protein function prediction from protein interaction network using physico-chemical properties of amino acids, *Int J Pharm Biol Sci* **4**:55–65, 2014.

24. Jiang JQ, McQuay LJ, Predicting protein function by multi-label correlated semi-supervised learning, *IEEE/ACM Trans Comput Biol Bioinform/IEEE, ACM* **9**:1059–1069, 2012.

25. Valentini G, Hierarchical ensemble methods for protein function prediction, *ISRN Bioinform* **2014**:1–34, 2014.

26. Zhang M-L, Zhou Z-H, A review on multi-label learning algorithms, *IEEE Trans Knowl Data Eng* **26**:1819–1837, 2014.

27. Yu G, Zhu H, Domeniconi C, Predicting protein functions using incomplete hierarchical labels, *BMC Bioinform* **16**:1, 2015.

28. Piovesan D, Giollo M, Leonardi E, Ferrari C, Tosatto, SCE, INGA: Protein function prediction combining interaction networks, domain assignments and sequence similarity, *Nucleic Acids Res* **43**:W134–W140, 2015.

29. Saha S, Sengupta K, Chatterjee P, Basu S, Nasipuri M, Analysis of protein targets in pathogen–host interaction in infectious diseases: A case study on Plasmodium falciparum and Homo sapiens interaction network, *Brief Funct Genomics*, doi: 10.1093/bfgp/elx024, 2017.

30. Saha S, Chatterjee P, Basu S, Nasipuri M, Gene ontology based function prediction of human protein using protein sequence and neighborhood property of PPI network, *Proc 5th Int Conf Front Intell Comput Theory Appl*, pp. 109–118, Springer, Singapore, 2017, doi: 10.1007/978-981-10-3156-4_11.

31. Saha S, Chatterjee P, Basu S, Nasipuri M, Functional group prediction of un-annotated protein by exploiting its neighborhood analysis in saccharomyces cerevisiae protein interaction network, in Chakri R, Saeed K, Cortesi A, Chaki N (eds.), *Advances in Intelligent Systems and Computing*, Springer, Singapore, 2017, pp. 165–177, doi: 10.1007/978-981-10-3391-9_11.

32. Prasad A, Saha S, Chatterjee P, Basu S, Nasipuri M, Protein function prediction from protein interaction network using bottom-up L2L Apriori algorithm, in Computational Intelligence Communications, and Business Analysis, pp. 3–16 (Springer, Singapore, 2017), doi:10.1007/978-981-10-6430-2_1.

33. Wang S, Wu F, Detecting overlapping protein complexes in PPI networks based on robustness, *Proteome Sci.* **11**:S18, 2013.

34. Singh M, Wadhwa PK, Kaur S, Predicting protein function using decision tree, *World Acad Sci Eng Technol* **2**:300–303, 2008.

35. Chen T, Guestrin C, XGBoost: A scalable tree boosting system, *Proc 22nd ACM SIGKDD Int Conf Knowledge Discovery and Data Mining*, pp. 785–794, 2016, doi: 10.1145/2939672.2939785.

36. Breiman L, Random forests, *Mach Learn* **45**:5–32, 2001.

37. Geurts P, Ernst D, Wehenkel L, Extremely randomized trees, *Mach Learn* **63**:3–42, 2006.

38. Pedregosa F, Varoquaux G, Gramfort A, Michel V, Thirion B, Grisel O, Blondel M, Prettenhofer P, Weiss R, Dubourg V, Vanderplas J, Passos A, Cournapeau D, Brucher M, Perrot M, Duchesnay E, Scikit-learn: Machine learning in python, *J Mach Learn Res* **12**:2825–2830, 2011.

39. Sengupta K, Saha S, Chatterjee P, Kundu M, Nasipuri M, Basu S, Ranked Gene Ontology based protein function prediction by analysis of protein-protein interactions, *Proc 6th Int Conf FICTA* (Springer, Nature, 2017).

40. Yousef A, Charkari NM, A novel method based on physicochemical properties of amino acids and one class classification algorithm for disease gene identification, *J Biomed Inform* **56**:300–306, 2015.

41. Li, X, Kierczak M, Shen X, Ahsan M, Carlborg O, Marklund S, PASE: A novel method for functional prediction of amino acid substitutions based on physicochemical properties, *Front Genet* **4**:1–6, 2013.

42. The UniProt Consortium. UniProt: A hub for protein information, *Nucl Acids Res* **43**:D204–D212, 2014.

43. Zhang, Lin H, Sang S, A method for predicting protein complex in dynamic PPI networks, *BMC Bioinform* **17**:229, 2016.

44. Chua HN, Sung W-K, Wong L, Exploiting indirect neighbours and topological weight to predict protein function from protein-protein interactions, *Bioinform (Oxford, England)* **22**:1623–1630, 2006.

45. Cozzetto D, Minneci F, Currant H, Jones DT, FFPred 3: Feature-based function prediction for all gene ontology domains, *Sci Rep* **6**:31865, 2016.

46. Kulmanov M, Khan MA, Hoehndorf R, DeepGO: Predicting protein functions from sequence and interactions using a deep ontology-aware classifier, *Bioinform* **34**:660–668, 2018.
47. Chatterjee P, Basu S, Kundu M, Nasipuri M, Plewczynski D, PPI_SVM: Prediction of protein-protein interactions using machine learning, domain-domain affinities and frequency tables, *Cell Mol Biol Lett* **16**:264–278, 2011.
48. Chatterjee P, Basu S, Zubek J, Kundu M, Nasipuri M, Plewczynski D, PDP-CON: Prediction of domain/linker residues in protein sequences using a consensus approach, *J Mol Model* **22**:72, 2016.
49. Chatterjee P, Basu S, Kundu M, Nasipuri M, Plewczynski D, PSP_MCSVM: Brainstorming consensus prediction of protein secondary structures using two-stage multiclass support vector machines, *J Mol Model* **17**:2191–2201, 2011.

Sovan Saha is an Assistant Professor in the Department of Computer Science and Engineering, Dr. Sudhir Chandra Sur Degree Engineering College, Dumdum, Kolkata, India since 2012. He received his Bachelor of Technology (Engineering) in Computer Science and Engineering from Saroj Mohan Institute of Technology, India, in 2010. He received his Master of Technology (Engineering) in Computer Science and Engineering from Netaji Subhash Engineering College, India, in 2012. His current research interest includes bioinformatics. He has published around 12 research articles in International Journals, Conference proceedings and book chapters in areas of Bioinformatics.

Abhimanyu Prasad completed his Bachelor of Technology (Engineering) in Computer Science and Engineering from Dr. Sudhir Chandra Sur Degree Engineering College, Dumdum, Kolkata, in 2017. His current research interest includes bioinformatics. He has published around three research articles in International Journals, Conference proceedings in areas of Bioinformatics.

Piyali Chatterjee is an Assistant Professor in the Department of Computer Science and Engineering, Netaji Subhash Engineering College, Kolkata, India since 2004. She received her Bachelor of Science degree in Computer Science and Engineering from Calcutta

University, India, in 2000. She received her Master of Science in Computer and Information Science from the same in 2002 and Master of Technology in Computer Science & Engineering from Calcutta University, 2004. She did Ph.D. in Engineering degree from Jadavpur University in 2012. Her current research interest includes bioinformatics and pattern recognition. Dr. Chatterjee has published around 25 research papers in various International Journals, Conference proceedings and book chapters in the areas of bioinformatics.

Subhadip Basu is a Professor in the Department of Computer Science and Engineering, Jadavpur University, Kolkata, India. He received his Bachelor of Engineering degree in Computer Science and Engineering from Kuvempu University, India in 1999 and Ph.D. (Engineering) degree from Jadavpur University in 2006. Professor Basu has published around 200 research articles in the areas of Bioinformatics, Biomedical Image Analysis, Character Recognition etc. He has co-authored/edited five books and co-invented two US patents. He is the recipient of the DAAD Fellowship from Germany, Research Award from UGC, Govt. of India, BOYSCAST and FASTTRACK young scientist fellowships from DST, Govt. of India, HIVIP fellowship from Hitachi, Japan and EMMA fellowships from the European Union. He is a senior member of IEEE and life member of IUPRAI, India.

Mita Nasipuri received her B.E.Tel.E., M.E.Tel.E, and Ph.D. (Eng.) degrees from Jadavpur University, in 1979, 1981 and 1990, respectively. Professor Nasipuri has been a faculty member of J.U since 1987. Her current research interest includes bioinformatics, image processing, pattern recognition, and multimedia systems. She has published around 450 research articles in various International Journals, Conference proceedings and book chapters in the areas of Bioinformatics, Pattern Recognition, Computer vision, Face Recognition, Classification etc. She is a senior member of the IEEE, U.S.A., Fellow of I.E. (India) and W.B.A.S.T, Kolkata, India.

CHAPTER 10

Protein Function Prediction from Dynamic Protein Interaction Network Using Gene Expression Data

Sovan Saha[*,§], Abhimanyu Prasad[*,¶], Piyali Chatterjee[†,||],
Subhadip Basu[‡,**] and Mita Nasipuri[†,††]

*Department of Computer Science & Engineering
Dr. Sudhir Chandra Sur Degree Engineering College
540, Dum Dum Road, Near Dum Dum Jn. Station, Surermath
Kolkata 700074, India*

†*Department of Computer Science & Engineering
Netaji Subhash Engineering College
Techno City, Panchpota, Garia
Kolkata 700152, India*

‡*Department of Computer Science & Engineering
Jadavpur University
188, Raja S.C. Mallick Road, Kolkata 700032, India*
§*sovansaha12@gmail.com*
¶*prasad.abhi150@gmail.com*
||*chatterjee_piyali@yahoo.com*
***bsubhadip@gmail.com*
††*mitanasipuri@yahoo.com*

Computational prediction of functional annotation of proteins is an uphill task. There is an ever increasing gap between functional characterization of protein sequences and deluge of protein sequences generated by

[a]This article was previously published in *Journal of Bioinformatics and Computational Biology*. Vol: 17, No. 4 (2019), 1950025 (15 pages).
§Corresponding author.

large-scale sequencing projects. The dynamic nature of protein interactions is frequently observed which is mostly influenced by any new change of state or change in stimuli. Functional characterization of proteins can be inferred from their interactions with each other, which is dynamic in nature. In this work, we have used a dynamic protein–protein interaction network (PPIN), time course gene expression data and protein sequence information for prediction of functional annotation of proteins. During progression of a particular function, it has also been observed that not all the proteins are active at all time points. For unannotated active proteins, our proposed methodology explores the dynamic PPIN consisting of level-1 and level-2 neighboring proteins at different time points, filtered by Damerau–Levenshtein edit distance to estimate the similarity between two protein sequences and coefficient variation methods to assess the strength of an edge in a network. Finally, from the filtered dynamic PPIN, at each time point, functional annotations of the level-2 proteins are assigned to the unknown and unannotated active proteins through the level-1 neighbor, following a bottom-up strategy. Our proposed methodology achieves an average precision, recall and F-Score of 0.59, 0.76 and 0.61 respectively, which is significantly higher than the reported *state-of-the-art* methods.

Keywords: Protein function prediction; dynamic protein interaction network; gene expression data; protein–protein interaction network.

1. Introduction

The use of high throughput technology results in the generation of a large number of proteins from sequenced genomes. But the functions of most of these proteins are not yet known. The conventional experimental procedures used for the detection of the functions of these unknown proteins need much more time, labor as well as cost. This fact necessitates the computational techniques to label functions of unknown proteins. Functional group prediction is a challenging computational problem as it is characterized by the following features: firstly it has to deal with a large number of functional groups. Secondly, a protein may be associated with more than one functions. Thirdly, the functional groups are related in hierarchy and fourthly, the functional groups in some proteins are in different proportion or uncertain or incomplete.

Protein function prediction problem can be solved in many ways: by sequence homology, phylogenetic profiles, domain information, protein complex, gene expression data, protein–protein interaction data, etc. Schwikowski[1] proposed a simple neighborhood counting method for protein function prediction which is mainly focused on the maximum number of occurrences of each functional group in the neighborhood graph of the unannotated protein. While Chen *et al.*[2] brought in a new concept of function similarity between the unannotated and its corresponding level-1 and level-2 neighbors based upon which the allocation of functional groups having higher similarity score to the unknown protein is done. Vazquez *et al.*[3] used simulated annealing methodology to generate an optimization problem which maximizes unknown protein connectivity by increasing the number of interconnected edges with it. Some other approaches like markov random field,[4] flow-based approach,[5] probabilistic methods,[6] network-based statistical algorithm,[7] binomial model-based loopy belief propagation,[8] bi-clustering-based UVCLUSTER[9] algorithm, graph clustering approaches[10,11] and Molecular Complex Detection (MCODE)[12] also emerge as successful and remarkable methodologies for protein–protein interaction network (PPIN), based protein function prediction. King *et al.*[13] proposed application of Restricted Neighborhood Search Clustering algorithm (RNCS) for clustering the PPIN which also serves as a beneficial information in forming PPIN for unannotated protein function prediction. Wei *et al.*[14] enriched the original PPIN by incorporating two types of edges: implicit and explicit to boost up the function prediction.

Combining multiple data sources to achieve the same goal empowers the prediction problem and lowers the risk of getting inaccurate results. It is known that inferences about functions can be made via protein–protein interaction, sequence homology but direct implications get hampered due to environmental change and dynamic nature of interactions. High throughput techniques generated PPIN may have erroneous data resulting into deviation of the correct prediction. Integration of protein–protein interaction data, time course gene expression data along with protein sequence similarity score and other derived neighborhood features may generate more accurate results with respect to real time.

Instead of utilizing individual information sources, integration of heterogeneous sources can provide confident functional assignments. Cozetto *et al.*[15] combines sequence, gene expression, protein–protein interaction data into a single framework which assigns functional categories probabilistically by analyzing complementary homology and features using Gene Ontology hierarchical structure. They propose novel scoring function COGIC for the evaluation of the semantic similarity between functional annotations.

The work of Peng *et al.*[16] attempts to use protein domain and protein complexes in PPIN. They not only incorporate domain combination similarity (DCS) of proteins and their neighbors but also consider DCS in the context of protein complexes (DSCP) separately. The proper integration of PPIN, protein's domain information and protein complexes could improve the accuracy.

Among other integrated approach, the work of Cao and Cheng[17] is worth mentioning. They take three probabilistic scores (MIS, SEQ, and NET score) to combine protein sequence, function association, protein–protein interaction and spatial gene–gene interaction. Finally, Statistical Multiple Integrative Scoring System (SMISS) is used to take integrated decision for protein function.

These computational methods considered PPIN as static entity. The fact is that proteins and their interactions are intrinsically controlled by different regulatory mechanisms through time and space. Recently, use of dynamic network has become a point of attraction for the research community.

In PPIN, interactions between proteins may vary in protein complexes and functional modules.[18] Protein complexes are groups of proteins that interact with each other at the same time and place such as Ap-2 adaptor complex, DNA polymerase epsilon complex, SAS complex, etc. Functional modules, consists of proteins that participate in a particular cellular process while binding each other at a different time and place, such as the CDK/cyclin module for cell cycle progression, etc.

Li *et al.*[19] distinguishes protein complexes and functional modules by integrating gene expression data into protein–protein interaction. In the work of Zhao *et al.*[20] a dynamic weighted PPIN is constructed from protein–protein interaction, time course gene expression data as well as protein's domain information.

PPI data collected from high throughput experiments might have false positives and false negative data. So, refinement of PPIN, incorporation of dynamic information of PPIN, level of sequence similarity between proteins of interest and their level-1 and level-2 neighbors, coefficient variation (CV) for each pair protein in the obtained PPIN may strongly predict the unknown functional group of candidate proteins. Considering this fact, a novel method using protein–protein interaction information and time course gene expression data is proposed where most active proteins are identified and taken as target proteins whose functional groups are to be predicted. Next, for those selected targets, their level-2 dynamic neighborhood graphs are considered. Finally using sequence similarity score and coefficient variation the appropriate functional groups are propagated from level-2 neighbors to targets via level-1 neighbors using bottom–up approach (see supplementary). Besides detailed implementation of Damerau–Levenshtein edit distance (DLD) methodology[21] (equation SE1 in supplementary), CV[22] (equation SE2 in supplementary) along with selection of active proteins in the dynamic neighborhood Graph in our proposed work has been discussed in supplementary document. The supplementary document is available online at https://drive.google.com/file/d/1eF0aFoiOw1aXt7hwxRb2QKVXrOqe86j4/view.

2. Dataset

The MIPS (Munich Information Center for Protein Sequences) database of yeast.[23] (ftp://ftpmips.helmholtzmuenchen.de/fungi/Saccharomycetes/CYGD/PPI/) has been used in this work for considering protein pair along with its corresponding functions. While for construction of dynamic PPIN, gene expression dataset of Spellman *et al.*[24] has been taken into account along with MIPS database. It actually contains gene names along with their corresponding scores (determined through experimental methods like micro-array, etc.) at various time points with an interval of 10 seconds. All the datasets are freely available online at https://www.dropbox.com/sh/chz454x6r5mmudk/AABKX-1DaCmFaK1Isvnbgusba?dl=0.

3. Methodology

Our proposed methodology has been categorized mainly into six steps, the flow diagram of which has been presented in Fig. 1. The detailed processing associated with each step has been described in the following sub-sections:

3.1. *Categorization of Proteins (According to Expression Level Over Time Points)*

In this work, different expression levels of genes or proteins at various sampling time points are considered. For a particular time point i, three

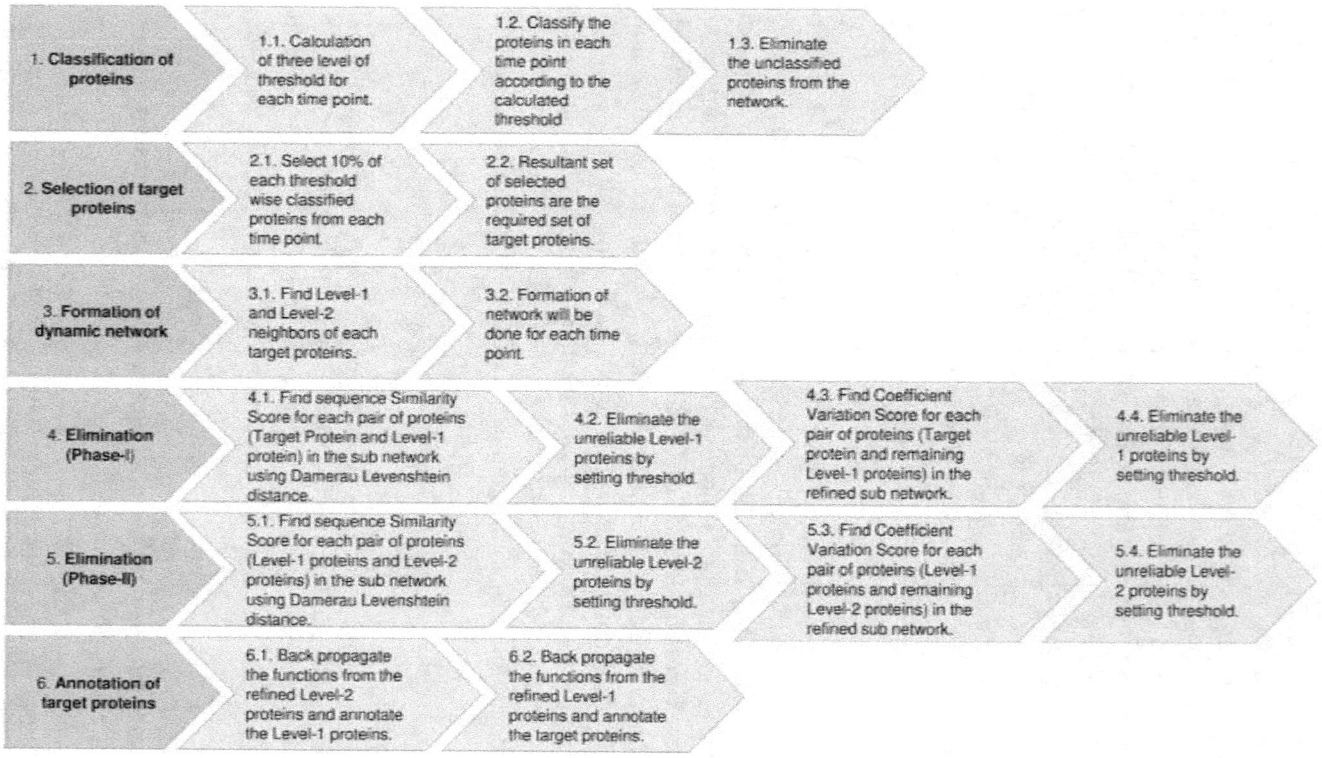

Fig. 1. **Flow Diagram of Our Proposed Methodology.** It consists of six stages: (1) Categorization of proteins (according to expression level over time points), (2) Selection of target proteins, (3) Formation of dynamic network, (4) Elimination (Phase-I), (5) Elimination (Phase-II), (6) Annotation of target proteins.

levels of thresholds (High, Medium, and Low) are taken. At each time point i. proteins expression levels are compared with thresholds and taken if they are within the range. Using different values of thresholds, reliable proteins with expression values are considered from time course gene expression data. The threshold values can be defined by equation SE3 in supplementary (see Fig. 2).

3.2. *Selection of Target Proteins*

Once the reliable proteins at a particular time point i with expression level at any of the three levels of thresholds (High, Medium, and Low) get selected, top 10% of proteins (most active proteins having high gene expression level) are taken from each of these three threshold levels at that particular time point. These top 10% of proteins (see Fig. 2) are considered as target protein set. The active probability (see supplementary) of all these selected proteins is also calculated (equation SE4–SE7 in supplementary).

3.3. *Formation of Dynamic Network*

For each unannotated target protein, its corresponding dynamic PPIN, at a particular time point, is formed using level-1 and level-2 proteins (see Fig. S4 in supplementary). Bottom-up approach (Fig. S1 in supplementary) is followed here. The level-1 proteins in this approach are considered as unannotated while only the level-2 neighbors or the leaf nodes are considered as annotated. So the functions of level-2 proteins are transmitted to the target protein through level-1 proteins.

3.4. *Elimination (Phase-I)*

Once the dynamic PPIN is formed, sequence similarity score $\mathrm{Sim}(i,j)$ using DLD metric is computed for each pair of target and its corresponding level-1 proteins. Proteins (level-1) having score below a particular threshold of sequence similarity score are eliminated. Then Coefficient Variation $(\mathrm{CV}(P_{U_i}^k, P_{A_i}^{\prime k}))$ has been computed between the target proteins and the remaining refined level-1 proteins. Proteins (level-1) having score below a particular threshold of CV score are eliminated. This phase-I of elimination ensures the non-availability of non-essential proteins in the PPIN in level-1 of unannotated target protein (Figs. S4 and S5).

Classification and Selection of Target Proteins

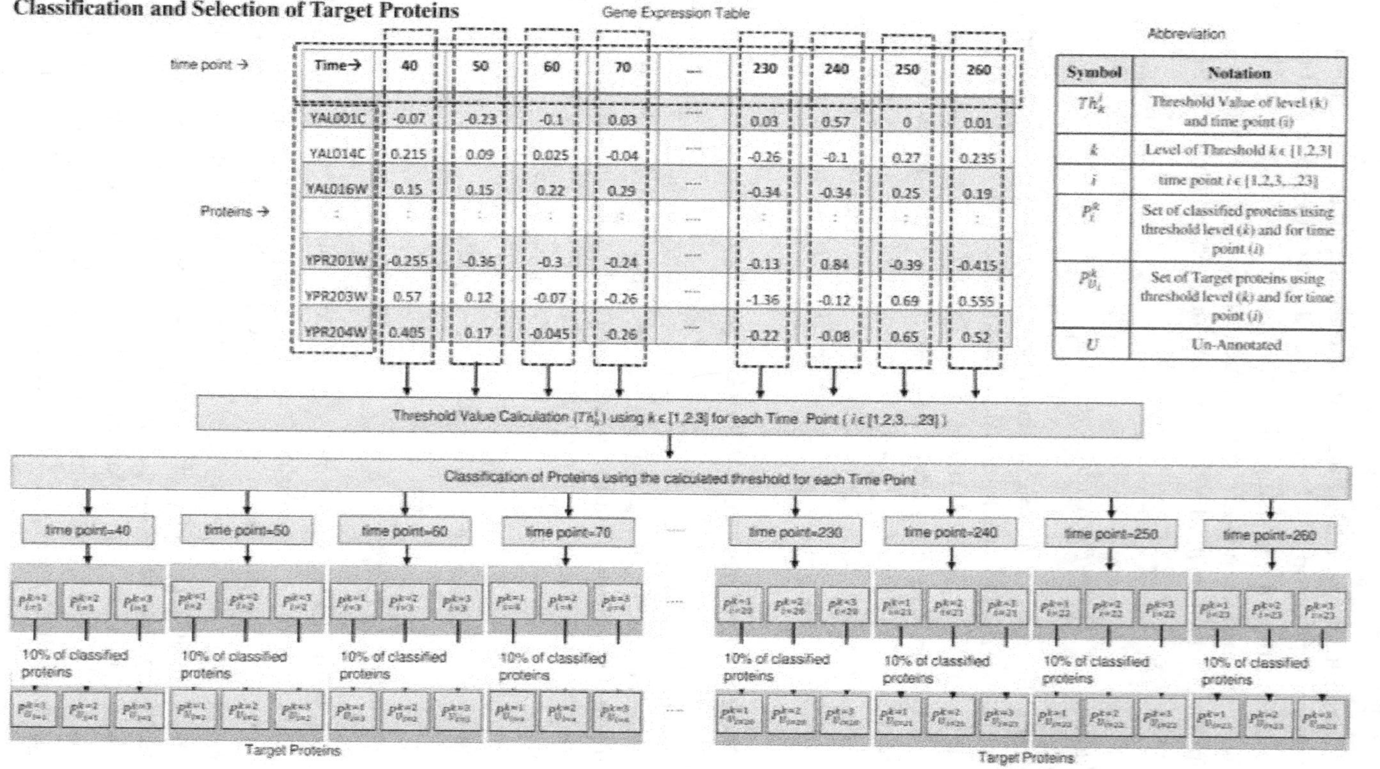

Time→	40	50	60	70	----	230	240	250	260
YAL001C	-0.07	-0.23	-0.1	0.03	----	0.03	0.57	0	0.01
YAL014C	0.215	0.09	0.025	-0.04	----	-0.26	-0.1	0.27	0.235
YAL016W	0.15	0.15	0.22	0.29	----	-0.34	-0.34	0.25	0.19
⋮	⋮	⋮	⋮	⋮	----	⋮	⋮	⋮	⋮
YPR201W	-0.255	-0.36	-0.3	-0.24	----	-0.13	0.84	-0.39	-0.415
YPR203W	0.57	0.12	-0.07	-0.26	----	-1.36	-0.12	0.69	0.555
YPR204W	0.405	0.17	-0.045	-0.26	----	-0.22	-0.08	0.65	0.52

Abbreviation

Symbol	Notation
Th_k^i	Threshold Value of level (k) and time point (i)
k	Level of Threshold $k \in [1,2,3]$
i	time point $i \in [1,2,3,...23]$
P_i^k	Set of classified proteins using threshold level (k) and for time point (i)
$P_{U_i}^k$	Set of Target proteins using threshold level (k) and for time point (i)
U	Un-Annotated

Fig. 2. **Categorization and selection of target proteins.** Representation of gene expression data that can be defined for various proteins at different time points followed by the selection of target set of active proteins at three threshold levels (High, Medium, and Low).

3.5. *Elimination (Phase-II)*

Once the refined dynamic PPIN is formed after Phase-I, sequence similarity score $\text{Sim}(i,j)$ using DLD metric is computed for each pair of level-1 and its corresponding level-2 proteins of target unannotated protein. Protein pairs (level-1 and level-2) having score below a particular threshold of sequence similarity score are eliminated. Then Coefficient Variation $(\text{CV}(P^k_{U_i}, P'^k_{A_i}))$ has been computed between the remaining refined level-1 and level-2 proteins. Protein pairs (level-1 and level-2) having score below a particular threshold of CV score are eliminated. This phase-II of elimination ensures the non-availability of non-essential proteins in the PPIN in level-1 and level-2 of unannotated target protein even after the execution of Phase-I. (Figs. S6 and S7). Thus, double filtering has been executed.

3.6. *Annotation of Target Proteins*

Finally, after the execution of double filtering, the ultimate refined PPIN is obtained from which functions of unannotated target proteins are predicted through the back propagation (Bottom-Up approach) of the functions of leaf nodes i.e. level-2 neighbors to the intermediate nodes, i.e. the unannotated level-1 neighbors and then transmitting the functions of the level-1 proteins to the unannotated target protein. (Figs. 3 and 4).

4. Results

Initially, our proposed methodology involves in the selection of active proteins at a particular time point at three levels of thresholds (High, Medium, and Low). Selection of target (10%) and training set (remaining 90%) from these active proteins at a particular time stamp at three levels of thresholds (High, Medium, and Low) is an important feature of our proposed methodology. Variation of total number of selected target proteins at various thresholds at sample time points has been highlighted in Table S1 in supplementary. The supplementary document is available online at https://drive.google.com/file/d/1eF0aFoiOw1aXt7hwxRb2 QKVXrOqe86j4/view.

Since dynamic PPIN is considered in our work, so a protein may be active at some time points while may be inactive at some others

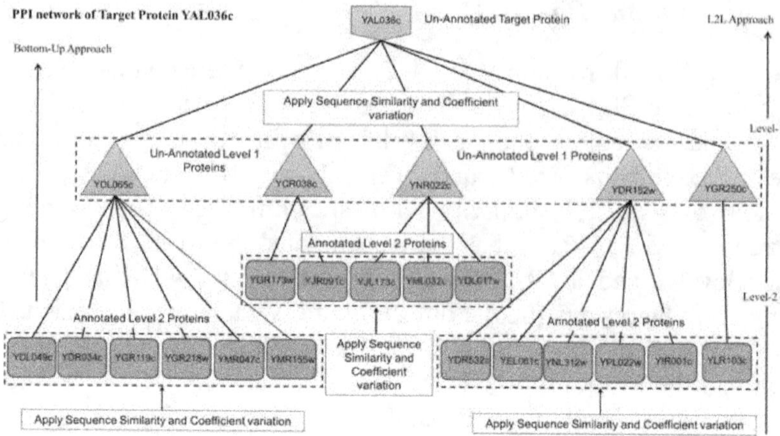

Fig. 3. **PPIN of target protein YAL036c.** Initial network of YAL036c before implementing our proposed methodology.

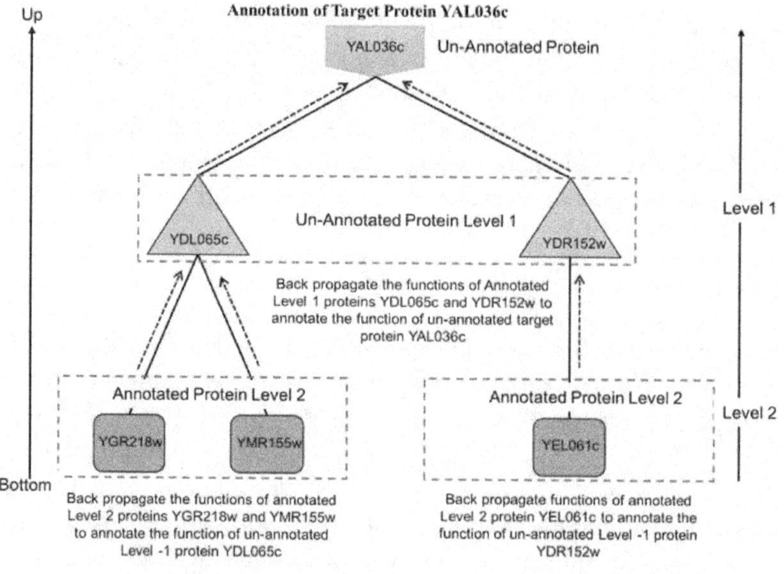

Fig. 4. **Annotation of Target Protein of YAL0036c.** Ultimate refined network of unannotated target protein YAL036c (original network shown in Fig. 3) after processing of our proposed methodology. Function of YAL036c is allocated following a Bottom-Up approach as described earlier.

(Table S2). In Table S2, target protein YBR160w is active at time points: 40, 50, 100, 120, 140, 190, and 230 while it is inactive at time points: 60, 70, 80, 90, 110, 130, 150, 160, 170, 180, 200, 210, 220, 240, 250, and 260. Similar cases have been observed for the other target proteins i.e. YCL029C, YDL074C, YCR009C, YDL017W, YCL027W, YBL033C, YBL061C, YBL101C, and YCL037C. Now, functions of the target protein at a particular time stamp can be predicted using our proposed methodology only if the target protein is active at that respective time point. Because it is not desirable to predict protein function which is inactive at a particular time point (Table S3 to S8 in supplementary).

From Table S2, the active time points of target protein YBR160W are shown. Since in Table S3, we have only considered four time points from 40 to 70, so it has been observed that YBR160W is active only at two time points out of the four, i.e. 40 and 50. So 33 functions of YBR160W has been predicted by our method at time stamp 40 out which all the functions get matched with the original 33 functions. While 35 functions of YBR160W has been predicted by our method at time point 50 out which 31 functions get matched with the original 33 functions. Similar instances have also been observed in the case of other target proteins (Table S3 to S8 in supplementary).

The activeness of selected proteins at three levels of thresholds (High, Medium, and Low) is very vital considering its presence in that interaction network. So the activeness is also verified for each selected protein at different conditions of thresholds (Eqs. SE4–SE7) at different time points. Some the samples are displayed in Fig. 5.

Moreover, measures are also taken to remove any non-essential proteins in level-1 and level-2 through the implementation of double filtering (by using sequence similarity score $\text{Sim}(i,j)$ using DLD metric and Coefficient Variation $(\text{CV}(P_{U_i}^k, P_{A_i}'^k))$ score) after the formation of dynamic network of each target protein. This is yet another vital aspect of our methodology along with the incorporation of dynamic network which makes it unique from the other existing protein function methodologies. The variation of the number of proteins in the pruned and original level-1 and level-2 neighbors of the unannotated target protein after double filtering at three different thresholds (High, Medium, and Low) at a particular time point is highlighted in Table S9.

Our methodology is evaluated by the standard measures: Precision, recall and F-Score. Variations of these standard measures at various

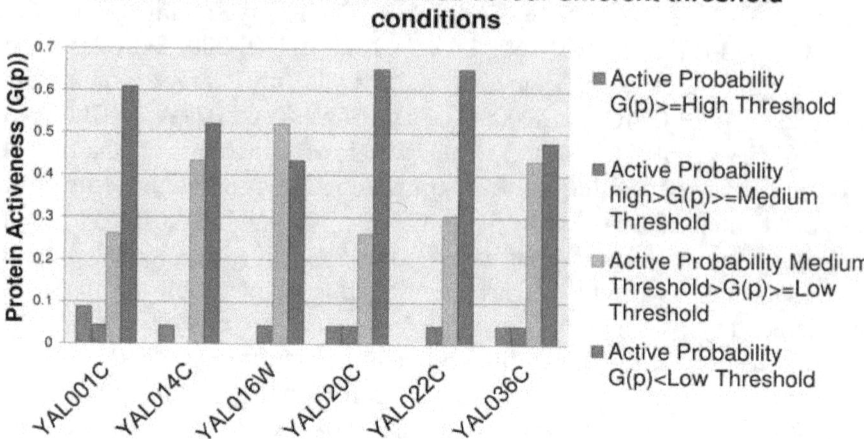

Fig. 5. **Protein activeness.** Bar Chart for the variation of protein activeness at different threshold conditions.

thresholds at different time points are displayed in Figs. S8–S10 in supplementary. It has achieved an overall precision, recall and F-score of 0.59, 0.76, and 0.61 respectively. It has been analyzed against other existing methodologies for unannotated protein function prediction like NRC,[30] FS-weight#1 (#1 and #2 represents level-1 and level-2 proteins),[29] Neighborhood counting#1(#1 represents only level-1 proteins),[29] Neighborhood counting#1 (#1 and #2 represents level-1 and level-2 proteins),[1] Chi square#1 (#1 represents only level-1 proteins),[28] Chi square#1 (#1 and #2 represents level-1 and level-2 proteins),[28] DeepGO[25] and ML_PFP.[26] It has been highlighted in Table 1. So, it is clearly indicated from Table 1, that our proposed methodology is the clear winner among all of them. Use of dynamic PPIN along with gene expression dataset and setting threshold at every phase helps this methodology to perform better eradicating the unreliable proteins from the network. Incorporation of CV taking degree of each node into consideration also helps in predicting functions of unannotated proteins from the denser sub graph or highly connected module in the PPIN. All these associate our proposed methodology to generate higher recall all throughout thereby generating less false positives in the prediction results.

Table 1. Comparison of our proposed methodology with other protein function prediction methodologies from PPIN like Chi-square method, Neighborhood Counting method, FS-weight method, DeepGO, ML_PFP, INGA, and NRC method.

Methods	Precision	Recall	F-score
Proposed Methodology	0.59	0.76	0.61
DeepGO[25]	0.03	0.07	0.04
ML_PFP[26]	0.51	0.58	0.54
INGA[27]	0.58	0.53	0.55
Chi square #1&2[28]	0.20	0.24	0.21
Chi square #1[28]	0.25	0.27	0.24
Neighborhood counting #1&2[1]	0.27	0.41	0.31
Neighborhood counting #1[1]	0.26	0.45	0.32
Fs-weight #1&2[29]	0.35	0.43	0.36
NRC[30]	0.37	0.43	0.36

So, in a nutshell, this prediction methodology definitely proves to be highly beneficiary in the field of medical science to predict the functions of unannotated proteins. Simultaneously, prediction of unannotated protein function using dynamic PPIN instead of static PPIN has uplifted this work to the next higher level since function predicted for an unannotated protein in a static PPIN may vary with respect to different time points in a dynamic PPIN. Moreover, two phase (Phase-I and Phase-II) filtering process helps to eliminate the non-essential proteins during the prediction of unannotated proteins in the dynamic PPIN both at level-1 and level-2. Two essential properties, i.e. protein sequence similarity and protein functional difference are considered during these two phase filtering process. The sequence similarity has been implemented by DLD metric while the functional difference is computed using CV. Domain[31-33] as well as functional categorization related information if combined with this dynamic PPIN based prediction might enhance its relevance in this field more effectively, which is our future work. And it should be added here that currently this work of dynamic PPIN is limited to only MIPS database of yeast which should be extended to other organisms as well as disease associated datasets like malaria, similar to our earlier work[34] with static PPIN.

Acknowledgments

Authors are thankful to the CMATER research laboratory of the Computer Science Department, Jadavpur University, India, for providing infrastructure facilities during progress of the work.

Funding

This project is partially supported by the CMATER research laboratory of the Computer Science and Engineering Department, Jadavpur University, India, and UGC Research Award (F.30-31/2016(SA-II)) from UGC, Government of India, and DBT project (No.BT/PR16356/BID/7/596/ 2016), Ministry of Science and Technology, Government of India.

References

1. Schwikowski B, Uetz P, Fields S, A network of protein–protein interactions in yeast, *Nat Biotechnol* **18**:1257–1261, 2000.
2. Chen J, Hsu W, Lee ML, Ng S-K, Labeling network motifs in protein interactomes for protein function prediction, *Proc Int Conf Data Engineering*, pp. 546–555, 2007.
3. Vazquez A, Flammini A, Maritan A, Vespignani A, Global protein function prediction from protein–protein interaction networks, *Nat Biotechnol* **21**:697–700, 2003.
4. Deng M, Mehta S, Sun F, Chen T, Inferring domain–domain interactions from protein–protein interactions, *Genome Res* **12**(10):1540–1548, 2002.
5. Nabieva E, Jim K, Agarwal A, Chazelle B, Singh M, Whole-proteome prediction of protein function via graph-theoretic analysis of interaction maps, *Bioinformatics* **21**(Suppl 1):i302–310, 2005.
6. Wu DD, An efficient approach to detect a protein community from a seed, *2005 IEEE Symp. Computational Intelligence in Bioinformatics and Computational Biology*, pp. 1–7, 2005.
7. Samanta MP, Liang S, Predicting protein functions from redundancies in large-scale protein interaction networks, *Proc Nat Acad Sci USA* **100**:12579–12583, 2003.
8. Letovsky S, Kasif S, Predicting protein function from protein/protein interaction data: A probabilistic approach, *Bioinformatics* **19**:i197–i204, 2003.
9. Arnau V, Mars S, Marín I, Iterative cluster analysis of protein interaction data, *Bioinformatics* **21**:364–378, 2005.

10. Altaf-Ul-Amin M, Shinbo Y, Mihara K, Kurokawa K, Kanaya S, Development and implementation of an algorithm for detection of protein complexes in large interaction networks, *BMC Bioinf* 7:207, 2006.

11. Spirin V, Mirny LA, Protein complexes and functional modules in molecular networks, *Proc Nat Acad Sci USA* **100**:12123–12128, 2003.

12. Bader GD, Hogue CWV, An automated method for finding molecular complexes in large protein interaction networks, *BMC Bioinf* **27**:1–27, 2003.

13. King AD, Przulj N, Jurisica I, Protein complex prediction via cost-based clustering, *Bioinformatics* **20**:3013–3020, 2004.

14. Xiong W, Liu H, Guan J, Zhou S, Sharan R, Ulitsky I, Shamir R, Sleator R, Walsh P, Altschul S, Madden T, Schäffer A, Zhang J, Zhang Z, Miller W, Lipman D, Friedberg I, Hulo N, Bairoch A, Bulliard V, Cerutti L, Wallace A, Laskowski R, Thornton J, Ye Y, Godzik A, Taubig H, Buchner A, Griebsch J, Wallace A, Laskowski R, Thornton J, Gilks W, Audit B, Angelis D d, Rost B, Liu J, Nair R, Schwikowski B, Uetz P, Fields S, Chua H, Wong L, Ng K, Ciou J, Huang C, Xiong W, Liu H, Guan J, Zhou S, Vazquez A, Flammini A, Maritan A, Karaoz U, Murali T, Letovsky S, Nabieva E, Jim K, Agarwal A, Chazelle B, Singh M, Deng M, Zhang K, Mehta S, Letovsky S, Kasif S, Kourmpetis Y, Dijk A v, Bink M, Brun C, Chevenet F, Martin D, Arnau V, Mars S, Marin I, Bu D, Zhao Y, Cai L, Dunn R, Dudbridge F, Sanderson C, Adamcsek B, Palla G, Farkas I, Derenyi I, Vicsek T, Becker E, Robisson B, Chapple C, Chua H, Sung W, Wong L, Hu L, Huang T, Shi X, Lu W, Cai Y, Sen P, Namata G, Bilgic M, Getoor L, Gallagher B, Eliassi-Rad T, Ashburner M, Catherine A, Judith A, Stark C, Breitkreutz B, Chatr-Aryamontri A, Ruepp A, Zollner A, Maier D, Albermann K, Güldener U, Münsterkötter M, Kastenmüller G, Strack N, Ruepp A, Doudieu O, Oever J v d, Brauner B, Damian S, Andrea F, Michael K, Milan S, Fan R, Lin C, Bogdanov P, Singh A, Protein function prediction by collective classification with explicit and implicit edges in protein–protein interaction networks, *BMC Bioinf* **14**:S4, 2013.

15. Cozzetto D, Buchan DW, Bryson K, Jones DT, Protein function prediction by massive integration of evolutionary analyses and multiple data sources, *BMC Bioinf* **14**:S1, 2013.

16. Peng W, Wang J, Cai J, Chen L, Li M, Wu F-X, Improving protein function prediction using domain and protein complexes in PPI networks, *BMC Systems Biology* **8**:35, 2014.

17. Cao R, Cheng J, Integrated protein function prediction by mining function associations, sequences, and protein–protein and gene–gene interaction networks, *Methods* **93**:84–91, 2016.

18. Spirin V, Mirny LA, Protein complexes and functional modules in molecular networks, *Proc Natl Acad Sci USA* **100**:12123–12128, 2003.

19. Li M, Wu X, Wang J, Pan Y, Towards the identification of protein complexes and functional modules by integrating PPI network and gene expression data, *BMC Bioinf* **13**:109, 2012.

20. Zhao B, Wang J, Member S, Li M, Li X, Li Y, Wu F-X, Pan Y, A new method for predicting protein functions from dynamic weighted interactome networks, *IEEE Trans Nanobiosci* **15**:131–139, 2016.

21. Bard GV, Spelling-error tolerant, order-independent pass-phrases via the Damerau–Levenshtein string-edit distance metric, *Conf Res Practice Inf Technol Ser* **68**:117–124, 2007.

22. Teng Z, Guo M, Liu X, Tian Z, Che K, Revealing protein functions based on relationships of interacting proteins and GO terms, *J Biomed Semantics* **8**(1):27, 2017.

23. Mewes H-W, Frishman D, Güldener U, Mannhaupt G, Mayer K, Mokrejs M, Morgenstern B, Münsterkötter M, Rudd S, Weil B, MIPS: A database for genomes and protein sequences, *Nucl Acid Res* **30**:31–34, 2002.

24. Spellman PT, Sherlock G, Zhang MQ, Vishwanath R, Anders K, Eisen MB, Brown PO, Futcher B, Comprehensive identification of cell cycle — regulated genes of the yeast saccharomyces cerevisiae by microarray hybridization, *Mol Biol Cell* **9**:3273–3297, 2003.

25. Kulmanov M, Khan MA, Hoehndorf R, Wren J, DeepGO: Predicting protein functions from sequence and interactions using a deep ontology-aware classifier, *Bioinformatics* **34**:660–668, 2018.

26. Saha S, Prasad A, Chatterjee P, Basu S, Nasipuri M, Protein function prediction from protein–protein interaction network using gene ontology based neighborhood analysis and physico-chemical features, *J Bioinf Comput Biol* **16**(6):1850025-1–1850025-15, 2018.

27. Piovesan D, Giollo M, Leonardi E, Ferrari C, Tosatto SCE, INGA: Protein function prediction combining interaction networks, domain assignments and sequence similarity, *Nucl Acids Res* **43**:W134–140, 2015.

28. Hishigaki H, Nakai K, Ono T, Tanigami A, Takagi T, Assessment of prediction accuracy of protein function from protein–protein interaction data, *Yeast* **18**:523–531, 2001.

29. Chua HN, Sung W-K, Wong L, Exploiting indirect neighbours and topological weight to predict protein function from protein–protein interactions, *Bioinformatics* **22**:1623–1630, 2006.

30. Moosavi S, Rahgozar M, Rahimi A, Protein function prediction using neighbor relativity in protein–protein interaction network, *Comput Biol Chem* **43**:11–16, 2013.

31. Chatterjee P, Basu S, Kundu M, Nasipuri M, Plewczynski D, PPI_SVM: Prediction of protein–protein interactions using machine learning, domain–domain affinities and frequency tables, *Cellular Mol Biol Lett* **16**:264–278, 2011.

32. Chatterjee P, Basu S, Zubek J, Kundu M, Nasipuri M, Plewczynski D, PDP-CON: Prediction of domain/linker residues in protein sequences using a consensus approach, *J Mol Model* **22**:72–87, 2016.
33. Chatterjee P, Basu S, Kundu M, Nasipuri M, Plewczynski D, PSP_MCSVM: Brainstorming consensus prediction of protein secondary structures using two-stage multiclass support vector machines, *J Mol Model* **17**:2191–2201, 2011.
34. Saha S, Sengupta K, Chatterjee P, Basu S, Nasipuri M, Analysis of protein targets in pathogen–host interaction in infectious diseases: A case study on Plasmodium falciparum and Homo sapiens interaction network, *Briefings in Functional Genomics* **17**(6):441–450, 2017.

Sovan Saha is an Assistant Professor in the Department of Computer Science and Engineering, Dr. Sudhir Chandra Sur Degree Engineering College, Dumdum, Kolkata, India since 2012. He received his Bachelor of Technology (Engineering) in Computer Science and Engineering from Saroj Mohan Institute of Technology, India, in 2010. He received his Master of Technology (Engineering) in Computer Science and Engineering from Netaji Subhash Engineering College, India, in 2012. His current research interest includes bioinformatics. He has published around 14 research articles in International Journals, Conference proceedings and book chapters in areas of Bioinformatics.

Abhimanyu Prasad completed his Bachelor of Technology (Engineering) in Computer Science and Engineering from Dr. Sudhir Chandra Sur Degree Engineering College, Dumdum, Kolkata, in 2017. His current research interest includes bioinformatics. He has published around 3 research articles in International Journals, Conference proceedings in areas of Bioinformatics.

Piyali Chatterjee is an Associate Professor in the Department of Computer Science and Engineering, Netaji Subhash Engineering College, Kolkata, India since 2004. She received her Bachelor of Science degree in Computer Science and Engineering from Calcutta University, India, in 2000. She received her Master of Science in Computer and Information Science from the same in 2002 and Master of Technology in Computer Science & Engineering from Calcutta University, 2004. She did Ph.D. in Engineering degree from Jadavpur University in 2012. Her current research interest includes bioinformatics, pattern recognition. Dr. Chatterjee has published around 25 research articles in various International Journals, Conference proceedings and book chapters in the areas of bioinformatics.

Subhadip Basu is a Professor in the Department of Computer Science and Engineering, Jadavpur University, Kolkata, India. He received his Bachelor of Engineering degree in Computer Science and Engineering from Kuvempu University, India in 1999 and Ph.D. (Engineering) degree from Jadavpur University in 2006. Prof. Basu has published around 200 research articles in the areas of Bioinformatics, Biomedical Image Analysis, Character Recognition, etc. He has co-authored/edited five books and co-invented two US patents. He is the recipient of the DAAD Fellowship from Germany, Research Award from UGC, Government of India, BOYSCAST and FASTTRACK young scientist fellowships from DST, Government of India, HIVIP fellowship from Hitachi, Japan and EMMA fellowships from the European Union. He is a senior member of IEEE and life member of IUPRAI, India.

Mita Nasipuri received her B.E.Tel.E., M.E.Tel.E, and Ph.D. (Engg.) degrees from Jadavpur University, in 1979, 1981, and 1990, respectively. Prof. Nasipuri has been a faculty member of J.U since 1987. Her current research interest includes bioinformatics, image processing, pattern recognition, and multimedia systems. She has published around 450 research articles in various International Journals, Conference proceedings and book chapters in the areas of Bioinformatics, Pattern Recognition, Computer vision, Face Recognition, Classification, etc. She is a senior member of the IEEE, U.S.A., Fellow of I. E. (India) and W.B.A.S.T, Kolkata, India.

CHAPTER 11

Explore the Hidden Treasure in Protein–Protein Interaction Networks — An Iterative Model for Predicting Protein Functions

Derui Wang* and Jingyu Hou†

School of Information Technology, Deakin University
221 Burwood Highway Burwood, Victoria 3125, Australia
deruiw@deakin.edu.au
†jingyu.hou@deakin.edu.au

Protein–protein interaction networks constructed by high throughput technologies provide opportunities for predicting protein functions. A lot of approaches and algorithms have been applied on PPI networks to predict functions of unannotated proteins over recent decades. However, most of existing algorithms and approaches do not consider unannotated proteins and their corresponding interactions in the prediction process. On the other hand, algorithms which make use of unannotated proteins have limited prediction performance. Moreover, current algorithms are usually one-off predictions. In this paper, we propose an iterative approach that utilizes unannotated proteins and their interactions in prediction. We conducted experiments to evaluate the performance and robustness of the proposed iterative approach. The iterative approach maximally improved the prediction performance by 50%–80% when there was a high proportion of unannotated neighborhood protein in the network. The iterative approach also showed robustness in various types of protein interaction network. Importantly, our iterative approach initially proposes an idea that iteratively incorporates the interaction information of

---

ᵃThis article was previously published in *Journal of Bioinformatics and Computational Biology*. Vol: 13, No. 5 (2015), 1550026 (22 pages).
*Corresponding author.

unannotated proteins into the protein function prediction and can be applied on existing prediction algorithms to improve prediction performance.

Keywords: PPI network; algorithm; protein function prediction; iterative model.

1. Background

Proteins are crucial biological molecules that play important roles in the expression of genes and assembly of other molecules. They also form the basic structure of organisms and act as vital parts in metabolism. Identifying protein functions is therefore a crucial task in molecule biology as it is necessary for studying biological processes. However, the cost of identifying protein functions through biological experiments could be relatively high and laboratory experiments could be inefficient in many circumstances. To solve these problems, predicting protein functions by computational approaches is becoming increasingly important.

In the early stage of *in silico* prediction methods, the most representative approach was based on the alignment of amino acid/nucleotide or the sequence motif searching (e.g. BLAST,[1] which is a widely used tool for protein function prediction through sequences searching and alignment). The idea behind using alignment to predict protein functions is that usually proteins having similar amino acid sequence tend to show the same functions in a biological pathway. However, sequence alignment methods lack efficiency due to the difficulty of determining sequence similarities between proteins. Moreover, it fails to consider diverse functions of protein.[2]

The analysis of protein functions in yeast (Saccharomyces cerevisiae) protein–protein interaction network (PPIN) showed that proteins having the same function and cellular location tend to be clustered together. It was discovered that 63% of the interactions occurred between proteins that have common functional assignment and 76% interactions occur between proteins that in the same subcellular compartment.[3] From the further research, it was found that 70%–80% of proteins share at least one function with their interacting partner.[4] On the other hand, about 70% proteins in yeast PPIN share functions with their level-1 proteins (the directly interacting partners with the protein in network) and level-2 proteins (the proteins directly interact with level-1 proteins).[5] As the consequence, a lot of mathematical and computational methods have been applied on PPINs of different species to predict functions of unannotated proteins.

A PPIN is denoted as an undirected graph $G(V, E)$, where V represents the vertices and E represents the set of edges in the network. Usually, PPI based methods for protein function prediction can be divided into two categories, which are direct methods and module-assisted methods.[6] For direct methods, at early stage, neighbor counting (NC) algorithm[3] and Chi-square algorithm[7] were proposed. These two simple unsupervised methods rely on statistical strategy to select the most frequent functions around unannotated proteins. Subsequently, function similarity weighted (FSW) algorithm[5] which utilizes network topology to assign weights to protein interactions was proposed. The Markov random field (MRF) approach proposed by Deng *et al.*[8] exploits function distributions in the PPIN to estimate parameters for function classification. Other machine learning algorithms, such as collective classification algorithm,[9,10] apply existing clustering methods on PPINs to cluster proteins and then protein function classifiers were derived based on function information in the clusters. There were also graphic theory based algorithms[11,12] and function flow (FF) method[13] which propagates annotations in PPINs to assign functions to unannotated proteins. As a trend of integrating multiple data sources for prediction, algorithms that incorporate semantic information of function annotations to predict protein functions were also extensively studied, such as the algorithms CIA[14] and DPA.[15] Furthermore, DCS/DCSP[16] integrates protein complexes and protein domain information along with PPIN to compute similarity between proteins and then assign functions to unannotated proteins based on protein similarity.

In module-assisted methods, discovering module is usually conducted by clustering similar proteins into a group. Then, unannotated proteins are annotated with the functions prevailing in the modules (i.e. groups) to which they belong. In a recent review,[17] graph-based clustering approaches were broadly categorized into five types: (i) Local neighborhood density search, such as MCODE,[18] DPClus,[19] and CFINDER.[20] Approaches in this category cluster nodes in PPINs based on density areas in PPINs; (ii) cost-based local search, such as RNSC[21] and OCG,[22] which performs local search to obtain optimum clustering in an effective way; (iii) flow simulation, such as MCL.[23] It employs random walk strategy on PPIN to build a Markov chain of transition matrixes, in order to discover clusters from transition matrix once it reaches convergence; (iv) link clustering, such as P_{EREIRA}[24] and AHN.[25] These methods cluster edges instead of nodes of the network; (v) population-based stochastic search, e.g. GA-PPI,[26] which uses genetic algorithm to generate clusters. Recently, some classic clustering methods were applied on integrated networks to discover functional modules in PPINs, for instance, SAMBA,[27] MS-KNN[28] and JCA.[29] These algorithms take advantages of associations among different types of biological data. Protein sequence, protein domain information and gene expression information are combined with PPINs to construct comprehensive networks in order to get better clustering performance.

Methods for integrating heterogeneous biological data were also studied, e.g. IWA,[30] to optimize the strategy for combining multiple data sources. Data sources are assigned weights based on their reliabilities in function and then networks built by each type of data are combined together as a weighed network that serves as a tool for prediction.

Currently, there are large proportions of unannotated proteins in the PPIN of each species (see Fig. 1).[6] However, a large proportion of prediction approaches that do not rely merely on PPIN topological information are sensitive toward unannotated proteins in the prediction domain (i.e. subnetwork that consists of proteins required to serve as the information sources for prediction algorithms or to participate in the prediction processes) since unannotated proteins, as well as their corresponding interaction information, do not participate in the prediction process. In other words, a lot of interaction information has been discarded when predicting functions, which affects the prediction performance significantly, especially for a high quality PPI dataset. This also means algorithms that utilize annotation information (e.g. semantic information-based methods or some classification algorithms) are incapable of predicting functions of a protein if most of its neighbor proteins are unannotated. On the other hand, the research in protein interactions in living cells discovered that proteins interact with each other, rather than working alone, to perform their functions in various biological processes.[2] Thus as a simulation of natural protein interacting process, it should be a more rational prediction model that functions of interacting proteins are decided mutually by both proteins in a bi-direction way. An iterative prediction approach based on iteratively updating protein similarities between annotated neighborhood proteins and the prediction target was

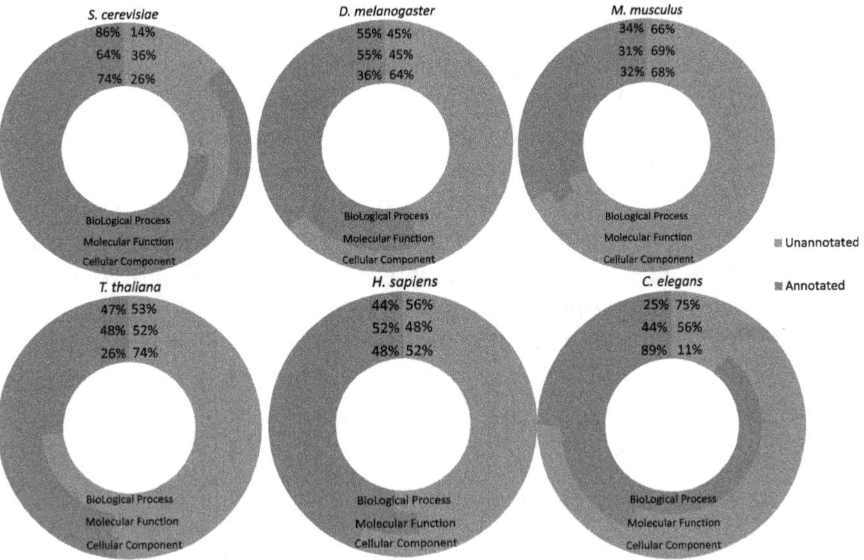

Fig. 1. The proportions of unannotated proteins in model species. (This graph shows the level of GO (see Refs. 31 and 34) annotations in some model species. Data used in this graph come from Rodeds survey paper (see Ref. 6)).

proposed[14] to address this problem. However, most of the current prediction algorithms do not reflect this feature of protein interactions and the prediction process was considered as a one-off process. That means the prediction is mono-directed from annotated proteins to unannotated proteins and once the functions of unannotated proteins have been predicted, the prediction is finished.

In this paper, we propose an iterative approach that makes use of the interaction information of unannotated proteins in the PPIN and iteratively updates the predicted functions of unannotated proteins in the network to avoid one-off prediction. The paper is organized as follows. In Sec. 2, we introduce our iterative approach. In Sec. 3, we present experimental results to compare the performance of our iterative approach with other reference algorithms and evaluate the robustness of our approach. Finally in Sec. 4, conclusions about the iterative approach and some future work are presented.

2. Iterative Model

Our iterative model is introduced in this section. Before prediction, we first need to find out unannotated proteins in the prediction domain. Then for each unannotated protein, its function will be initialized respectively by a prediction algorithm (e.g. NC algorithm or FSW algorithm in our case). After every unannotated protein in the prediction domain is initially annotated, predicted functions in the prediction domain are then iteratively updated until every unannotated protein gets its predicted functions that do not change anymore. The approach is divided into four steps: Unannotated protein searching, protein function initialization, iterative function updating and prediction result selection, which are illustrated in Fig. 2. The pseudo code of our iterative model is provided in the appendix.

2.1. *Unannotated Protein Searching*

In this first step, with a PPIN $G(VE)$, and the GO annotation file, unannotated proteins in the prediction domain (e.g. in NC algorithm, it is the directly interacting partners of protein to be predicted) around the main prediction target protein, P_x, are found out from the PPIN. Main prediction target P_x and other unannotated proteins are all stored in set ψ as the target proteins.

2.2. *Protein Function Initialization*

After finding out unannotated target proteins in the prediction domain of the main prediction target, each unannotated target protein's functions will be initialized. Suppose there are totally n proteins (including the main prediction target protein, P_x) in set ψ. For the function initialization of the ith protein P_i $(1 \leq i \leq n)$ in ψ, other unannotated proteins, φ $(\varphi = \psi - \{P_i\})$ and interactions associated with them are removed from the PPIN during the initialization. Then, functions of the unannotated target protein P_i are initially predicted using an existing protein function prediction algorithm. We then similarly initialize functions for all n proteins in ψ until functions of all n proteins have been initially predicted. For an unannotated target protein which has no annotated neighbors, its function is initialized as null (unannotated). The function initialization starts from a randomly selected protein in ψ and in a random order in our

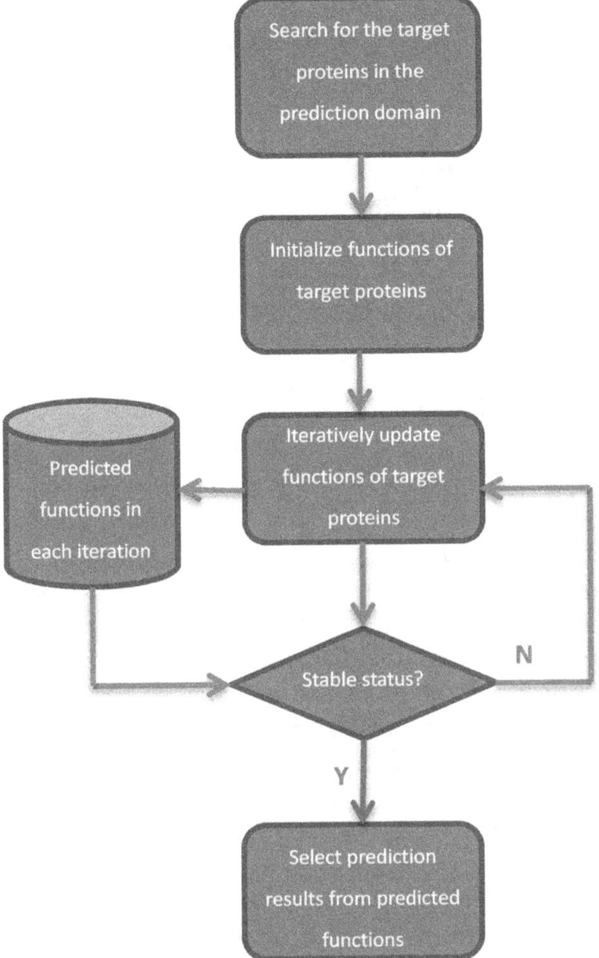

Fig. 2. Diagram of the proposed iterative approach.

approach. This initialization is regarded as the first round of prediction (the 1st iteration).

2.3. *Iterative Function Updating*

After function initialization, the target proteins with initialized non-null functions then join back into the prediction domain as annotated proteins at the beginning of each iteration. Therefore, functions of these proteins can be re-predicted and updated since the protein function distribution in the prediction domain is changed.

Details of the iterative function updating are as follows. (i) At the beginning of the tth iteration ($t \in [1, M], t = 1$ for the protein function initialization), a protein P_i ($1 \leq i \leq n$) is randomly selected from the set ψ, then a protein function prediction algorithm is applied to re-predict its functions. (ii) Repeat step (i) operation for all P_i in set ψ one by one, until functions of each P_i have been re-predicted and updated. Then, the current tth round of prediction ends and the $(t + 1)$th prediction starts. (iii) If re-predicted functions of P_i after the $(t + 1)$th round of prediction are different from predicted functions after tth round of prediction, then its functions are updated as the new re-predicted functions; otherwise keep the functions of P_i unchanged. (iv) Repeat steps (ii) and (iii) until functions in each protein reach a steady status after the Mth iteration (i.e. functions of every protein in ψ do not change any more or repeat in a pattern), then the iterative function updating ends. If the predicted functions of P_i repeat in a certain pattern after the Mth round of prediction and have a repeat span of T (i.e. prediction results from the $(t + T)$th prediction are the same as the predicted functions from the tth prediction when $t \geq M$), then we randomly choose one group of functions from these T groups of predicted functions as the prediction result of P_i. Meanwhile, we set M as the total number of iteration before reaching a stable status.

In each round of iteration, the target proteins for which the function prediction is conducted are selected in a random order. The robustness of the iterative function updating toward different orders of proteins being predicted is evaluated in the experiment section.

2.4. *Prediction Result Selecting*

After each round of iteration, the predicted functions of each target protein are recorded. When the iteration ends, for each protein, its final predicted functions are selected statistically from all its predicted functions generated from different rounds of iterations.

Let the set of whole function annotations in the PPIN be F and F $=$ $\{f^1, f^2, f^3, \ldots, f^K\}$, where f^1 to f^K are function annotations and K is the number of function annotation terms in the PPIN. For an unannotated protein P_x (the main prediction target), predicted functions after tth step of iteration are recorded as $F_{tP_x} = [f_{x,t}^1, f_{x,t}^2, \ldots, f_{x,t}^K]^T$, $f_{x,t}^j = 1$ if the predicted functions of protein P_x from the tth iteration contains function $f^j (f^j \in F)$, otherwise $f_{x,t}^j = 0$. F_{1P_x} records the initial functions of P_x. When the iteration reaches the stable status after the Mth iteration, we can have a matrix AF_{P_x} that records predicted functions of P_x generated from all steps of the iteration

$$AF_{P_x} = [F_{1P_x}, F_{2P_x}, F_{3P_x}, \ldots, F_{MP_x}].$$

F_{tP_x} is the vector containing predicted functions of protein P_x after the tth iteration. Then we have

$$C_{P_x} = AF_{P_x} \times I.$$

$I = [1, 1, \ldots, 1]^T$, $C_{P_x} = [C_{Px}^{f1}, C_{Px}^{f2}, C_{Px}^{f3}, \ldots, C_{Px}^{fK}]^T$ and $C_{Px}^{fj} = \sum_{t=1}^M f_{x,t}^j$ $(1 \leq j \leq K)$. The final predicted functions of the unannotated protein P_x are selected as follows:

$$\text{Predicted function} = \left\{ f^j | f^j \in F, C_{Px}^{fj} \geq \alpha \times \max_{(1 \leq j \leq K)} \left(C_{Px}^{fj} \right) \right\}, \quad \alpha \in [0, 1].$$

In our experiment, we set the parameter $\alpha = 0.5$. Actually, the predicted functions are selected based on the frequency of their occurrences in the whole iteration process. It can be seen that the one-off algorithms are the special case of this iterative model with the total number of iteration $M = 1$.

2.5. *Function Prediction Algorithms*

Here, we present three PPI based function prediction algorithms that can be used as the base/reference prediction algorithms in our iterative prediction model. These three algorithms are NC algorithm, FSW algorithm and FF algorithm. The NC algorithm was proposed by Schwikowski *et al.* in 2000. It predicts functions of each protein by counting the frequency of each function being found in its neighborhood proteins. The more frequent a function shows in the protein's level-1 neighbors (directly interacting proteins), the more likely it is a function of the protein. Let p stand for the unannotated protein to be predicted, N_p is the set of level-1 proteins of p, p' is a protein in N_p. $\delta(p', f) = 1$ if function f is found in p', otherwise $\delta(p', f) = 0$. A function f is scored based on the formula

$$\text{Score}(p,f) = \sum_{p' \in N_p} \delta(p',f).$$

A function with a higher score would be chosen as the prediction result. Originally, the NC algorithm chooses the top three functions as the functions of the unannotated protein p.

The FSW algorithm defines a similarity between proteins based on the number of their commonly interacting proteins in a PPIN, then assigns weights to proteins based on their similarity with the prediction target. Thus, a similarity among proteins that share their neighbors and highly connected should be relatively higher than others. Functions in those proteins that have higher similarities with the unannotated protein would get higher weights in function voting. Actually, let N_u and N_v denote level-1 neighbor protein sets of two proteins u and v, the similarity of these two proteins, $S(u,v)$, is given by

$$S(u,v) = \frac{2|N_u \cap N_v|}{|N_u - N_v| + 2|N_u \cap N_v|} + \frac{2|N_u \cap N_v|}{|N_v - N_u| + 2|N_u \cap N_v|}.$$

If protein u is similar to protein w and protein w is similar to protein v, then u and v have transitive functional association and similarity between u and v could be depicted as a transitive FSW similarity

$$S_{TR}(u,v) = \max(S(u,v),\ \max_{w \in Nu} S(u,w)S(w,v)).$$

$S(u,v)$ is the FSW between u and v. The score of a function f in level-1 proteins and level-2 proteins of the unannotated prediction target u is defined as

$$\text{Score}(u,f) = \sum_{v \in N_u} \left[S_{TR}(u,v)\delta(v,f) + \sum_{w \in N_v} S(u,w)\delta(w,f) \right].$$

$\delta(v,f)$ and $\delta(w,f)$ are indicator functions with $\delta(p,f) = 1$ if protein p has function f, otherwise $\delta(p,f) = 0$. The predicted functions are those with higher scores. The prediction domain of the NC algorithm is defined as the unannotated protein and its level-1 neighbor proteins, while for the FSW algorithm it is the unannotated protein and its level-1 and level-2 neighbor proteins.

Function flow[13] was used as another reference algorithm. Functions propagate in the network as flows. Annotated proteins are regarded as sources of functional flows (a reservoir). An iterative algorithm using discrete time is used to simulate the spread of functional flows in neighborhood proteins. After d iterations, the functional score of a protein corresponds to the total amount of functional flows that the protein has received during the iteration process.

For each protein u in the network, a variable $R_t^f(u)$ is defined corresponding to the amount in reservoirs for function f that node u has at time t. At time zero, there are only reservoir of function f at node u

$$R_0^f(u) = \begin{cases} \infty & \text{if } u \text{ is annotated with } f, \\ 0 & \text{otherwise.} \end{cases}$$

At each time step, functional flows proceeding downhill from node u to v satisfy the capacity constraints

$$g_t^a(u, v) = \begin{cases} 0 & \text{if } R_t^f(u) < R_{t-1}^f(u), \\ \min\left(w_{u,v}, \dfrac{w_{u,v}}{\sum_{(u,y)\in E} w_{u,y}} \right) & \text{otherwise.} \end{cases}$$

In the formula, $w_{u,v}$ is the weight of edge (u, v). $w_{u,v} = 1$ in unweighted PPIN and E is the set of edges in network. Reservoir of each protein is recomputed as there are functional flows entering and leaving the nodes.

$$R_t^f(u) = R_{t-1}^f(u) + \sum_{v:(u,v)\in E} (g_t^f(v, u) - g_t^f(u, v)).$$

Finally, the functional score for protein u and function f over d iterations is calculated as the total amount of flow that has entered the node. Functions with the highest score are selected as the predicted functions

$$\text{Score}(u,f) = \sum_{t=1}^{d} \sum_{v:(u,v)\in E} g_t^f(v, u).$$

3. Results and Discussion

3.1. *Datasets*

It is noted that PPIN usually contains noisy information due to a lot of false positive interactions generated from high through-put technologies.

In our experiments, evaluation of the algorithm performance on the networks with different qualities was conducted. Totally four datasets were employed for the experiments. We employed function category (FunCat) annotation[36] in our experiments. There are hierarchical annotations of 6167 proteins in the dataset. For the purposes of prediction performance evaluation, unannotated proteins in the original PPINs were removed.

The first dataset was the Munich information center for protein sequences (MIPS) yeast PPIN from FTP server of MIPS (MIPS: ftp:// ftpmips.gsf.de/fungi/yeast/PPI/). After removing 250 unannotated proteins, the dataset contains 4,554 proteins and 15,456 interactions. The MIPS dataset comprises physical interaction data including interactions collected from small-scale experiments and some core data generated from high throughput technologies,[32,33] and it is believed to be highly reliable. After downloading the raw data, we filtered redundant and self-interactions in the network. For the evaluation purpose, we removed unannotated proteins from the network. The MIPS dataset after pre-processing had 4,273 proteins and 12,735 interactions in it.

Another dataset was the BioGrid yeast PPIN downloaded from saccharomyces genome database (SGD: http://www.yeastgenome.org/). The BioGrid PPIN contains physical interactions and genetic interactions generated by high throughput technologies, thus the dataset contains a high proportion of false positive interactions. We filtered redundant and self-interactions in the network. For the evaluation purpose, we removed unannotated proteins from the network as well. To make its network scale comparable with the scale of the MIPS network, we randomly selected a subnetwork of BioGrid network. After filtering out 216 unannotated proteins, the final dataset contained 4,249 proteins and 10,000 interactions for our experiment.

The third dataset was the intersection of MIPS yeast network and BioGrid network (named as BioGrid–MIPS network). This dataset consists of 3,025 distinct annotated proteins and 8,405 interactions.

We also combined gene co-expression data of yeast from COX-PRESdb[35] (COXPRESdb: http://coxpresdb.jp/) into the original PPINs to further test the performance of the iterative model on the cases where the original PPIN structure is modified due to the incorporation of related genomic information. Data from COXPRESdb contains 4,461 expression information files generated from 3,819 gene chips, each file represents a gene and the corresponding co-expressed genes are listed as the file content. Co-expressed genes had already been processed and ranked from high to low based on the extent of co-expression. We constructed a gene co-expression network by adding an edge between a gene and its first ranked co-expressed gene. After a gene co-expression network was constructed, we modified the MIPS PPIN by adding edges between proteins that had the corresponding edges in the gene co-expression network i.e. we merged the co-expression network with MIPS network to form a new dataset.

3.2. *Performance Metrics*

We measured the performance of prediction in terms of precision and recall. Since there is a trade-off between precision and recall, we also used F-value to measure the performance of the prediction results. Actually, let N_t be the total number of true functions (generated from laboratory) in protein P, N be the number of predicted functions and N_c be the number of correctly predicted functions. Then precision, recall and F-value are defined as follows:

$$precision = \frac{N_c}{N}, \quad recall = \frac{N_c}{N_t},$$

$$F\text{-value} = \frac{2 \times precision \times recall}{precision + recall}.$$

Usually, higher precision, recall and F-value indicate a better performance of an algorithm. Also, we used precision–recall curves to compare precision at different levels of recall.

3.3. *Prediction Performance*

First, we evaluated the performance of the iterative model on proteins that have certain levels of unannotated neighbors. We compared the precision, the recall and the F-value of the iteration algorithms and the non-iteration algorithms (also known as reference algorithms) on four datasets which are the MIPS PPIN, the BioGrid PPIN, the intersection network of MIPS and BioGrid PPIN, and the Merged network of MIPS PPIN and COXPRESdb gene co-expression network.

We used leave-one-out cross (LOOC) validation method in the evaluation. For each of the above four datasets, we randomly selected nine groups of proteins as main prediction targets (i.e. the proteins for which we supposed their functions were unknown, thus we could apply prediction algorithms on them). Each group contained 100 randomly selected main prediction target proteins. From group 1 to group 9, in each dataset, 10% to 90% of proteins in the prediction domain of each target protein were randomly set as unannotated proteins. Then, we evaluated our iterative model and reference algorithms on these 36 protein groups (3,600 main prediction targets with different percentage of unannotated proteins in their prediction domain), respectively. The average precisions, the average recalls and the average F-values of each group were recorded for comparison.

In order to evaluate the performance of the proposed iterative model, we applied the iterative model on iterative neighbor counting (INC) algorithm and then compared INC with original NC algorithm. Furthermore, we compared our iterative approach with the FSW algorithm and FF algorithm which also utilize unannotated proteins and corresponding interactions in prediction processes. To achieve comparability among iterative and non-iterative approaches, we applied the idea from FSW to weigh protein pairs but excluded those unannotated proteins from the neighbor in each step of the iterative neighbor counting method, so that the topological structure of the neighbor and the predicted functions of the unannotated protein were updated iteratively. We named this iterative model iterative weighted neighbor counting (IWNC). To make the results comparable, we used unified prediction domain (level-1 and level-2 neighbors around unannotated nodes) for IWNC, FSW and FF algorithm.

We compared INC algorithm and NC algorithm on constructed four datasets. Then, we compared IWNC algorithm, FSW algorithm and FF method similarly. Comparisons of prediction results are presented in Fig. 3.

At first, we compared results from different algorithms on the same dataset. INC was compared with NC, while IWNC was compared with FSW and FF. The above experiment results showed that our iterative algorithms (i.e. INC and IWNC) outperformed the reference algorithms in terms of precision, recall and F-value in most cases. Especially, when the percentage of unannotated proteins in the prediction domain increased, precision, recall and F-value of NC algorithms showed rapid decrease, while the iterative model INC kept relatively remarkable prediction performance. Comparison between IWNC and FSW shows that IWNC

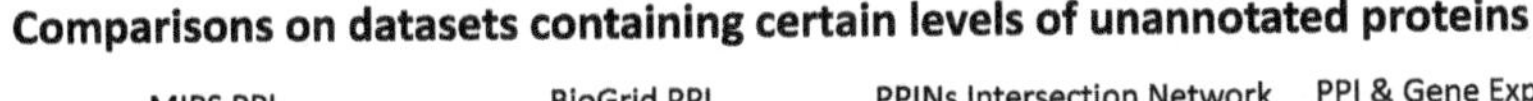

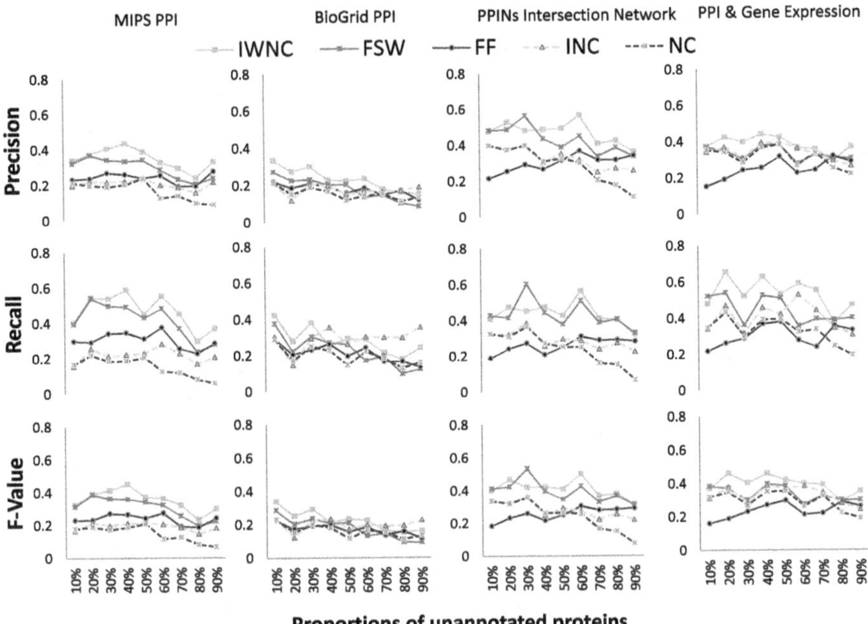

Fig. 3. Comparisons on datasets containing certain levels of unannotated proteins (results were from datasets consisting of certain proportion of unannotated proteins. Precisions, recalls and F-values from one dataset are placed in the same column).

generated better prediction results in most cases, even the unannotated proteins did not participate in computing of protein similarities during the prediction process. The similar conclusion could be drawn from the comparison between IWNC algorithm and FF method. Furthermore, the iterative approach kept robust F-values to the variation of unannotated protein numbers in the prediction domain and the gene co-expression data information.

Second, we compared results on different datasets. It could be observed that the algorithms generated better prediction results from the dataset that was the intersection of the MIPS and BioGrid datasets. That is because the protein interactions in this dataset are endorsed by both MIPS and BioGrid datasets, so they have a higher credibility. Correspondingly, the PPINs intersection dataset has less noisy information, which increased the prediction performance. From the results generated from BioGrid dataset, we observed that IWNC, FSW and FF method did not perform as good as they did on other datasets. This was possibly caused by the quality of selected BioGrid datasets. Since we sparsified the dataset to make its scale smaller than the original BioGrid network, it had a greater possibility that the noisy information in the dataset could affect the prediction results. However, results from MIPS datasets, intersection datasets of MIPS and BioGrid and MIPS data contain gene expression information showed that the IWNC algorithm performed better than the FSW algorithm and the FF method in most cases.

After evaluating the iterative model performance and robustness toward the increasing percentages of unannotated proteins in the prediction domain, we then evaluated the overall performance of the iterative model on the prediction domains with random proportions of unannotated proteins, which simulates the situation in real world PPINs. We chose dataset A: 10 groups of proteins from the MIPS dataset and dataset B: 10 groups of proteins from the BioGrid dataset. Each group contains 50 proteins that have random proportions of unannotated neighbors. Then, we tested performance of INC and NC, IWNC, FSW and FF algorithms on datasets A and B.

For dataset A, we recorded the precision, recall and F-value of each predicted protein P_i as precision$_i$ and recall$_i$ and F-value$_i$. We then calculated average precision and average recall values of each group of proteins as

$$\text{Avg(precision)}_{gA} = \frac{\sum_{i=1}^{50} \text{precision}_i}{50}, \quad g = 1, 2, 3, \ldots, 10,$$

$$\text{Avg(recall)}_{gA} = \frac{\sum_{i=1}^{50} \text{recall}_i}{50}, \quad g = 1, 2, 3, \ldots, 10,$$

$$\text{Avg(F-value)}_{gA} = \frac{\sum_{i=1}^{50} \text{F-value}_i}{50}, \quad g = 1, 2, 3, \ldots, 10.$$

In which g is the gth group of proteins in dataset A. So we obtained 10 pairs of precision–recall values (i.e. from $(\text{avg(precision)}_{1A}, \text{Avg(recall)}_{1A})$ to $(\text{Avg(precision)}_{10A}, \text{Avg(recall)}_{10A})$) for each algorithm. First, we plotted comparisons of $\text{Avg(precision)}_{gA}$, Avg(recall)_{gA} and Avg(F-value)_{gA} of each algorithm. Then, for each algorithm we plotted its precision–recall curve which consists of 10 pairs of precision–recall values as mentioned. The same procedures were conducted on dataset B. Finally, we present the comparisons of precision, recall, F-value and precision–recall curves of iterative approach and reference algorithms.

(i) Comparisons of INC and NC algorithm on two datasets (Fig. 4).
(ii) Comparisons of P–R Curves (Fig. 5).
It is shown from the above overall performance evaluation results that our iterative model surpassed reference algorithms in terms of precision, recall and F-value on both datasets.

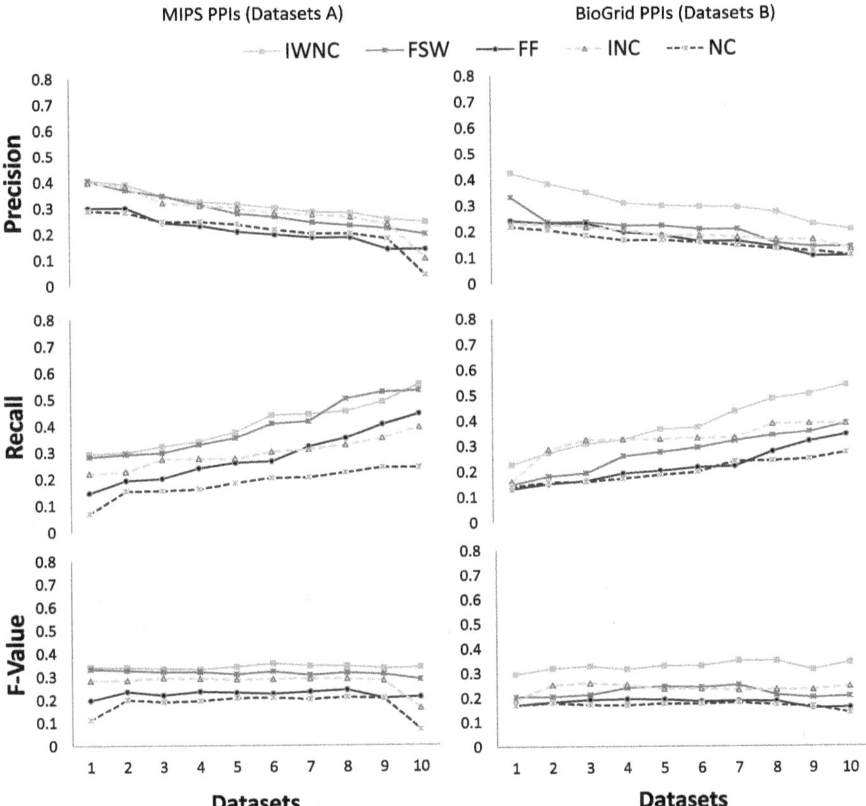

Fig. 4. Comparisons on datasets containing random levels of unannotated proteins (results were from datasets consisting of random proportion of unannotated proteins. Precisions, recalls and F-values from one dataset are placed in the same column).

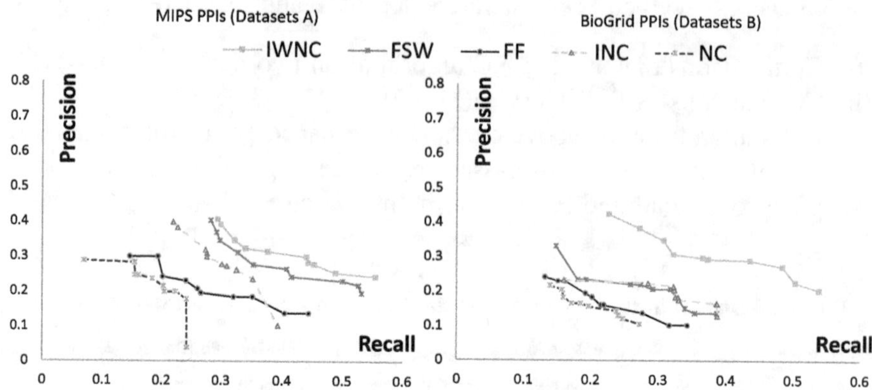

Fig. 5. P–R curves of iterative and non-iterative algorithms.

3.4. *Sensitivity Analysis Over Iteration Times*

When the iteration times increase, the importance of the initial prediction may decrease. Thus, a sensitivity analysis over iteration times before achieving a stable status was carried out to show the prediction performance with different times of iteration.

In order to reveal trends of sensitivity variation under datasets that consist of different rates of unannotated proteins, we chose three sets of proteins from MIPS network. The first set contained 100 proteins each of which has 10% unannotated neighbors. Another 100 proteins in the second dataset had random proportion of unannotated neighbors. In the third dataset, there are 100 proteins each of which has 90% unannotated neighbors. We then applied INC to predict functions of each protein. We recorded the times of iteration on each dataset and the corresponding true positive rate (i.e. recall) of the predicted results (see Fig. 6).

It can be observed that the iteration needs more steps to reach the stable status when the unannotated proteins are of a higher proportion. Statistically, in the same dataset, the predicted functions had higher true positive rates when the prediction went through more iterations.

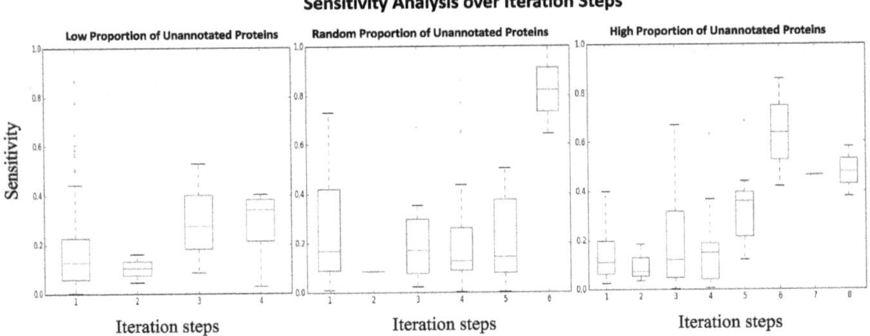

Fig. 6. Sensitivity analysis of INC on datasets which consist of different proportions of unannotated proteins.

3.5. *Robustness of the Iterative Approach*

As indicated in our algorithm, for each round of iteration, unannotated proteins were predicted in a random order. We randomly chose predicted results for those target proteins whose predicted functions repeated in a pattern after certain rounds of iteration. Thus, another issue we were concerned about is whether the final predicted functions will be changed from different runs of prediction (i.e. the robustness of the iterative model). We ran our algorithm on the MIPS network to evaluate the robustness of our iterative model to different prediction orders of the target proteins. During each run, the prediction started from different target proteins at the beginning of each iteration and different runs also had different orders of prediction. We then analyzed the prediction results from different runs. For instance, we set 50% of protein YHL011C's neighborhood proteins as unannotated. FunCat annotations of the predicted functions in different iteration orders from independent runs are listed in Table 1.

Furthermore, we randomly chose 100 proteins that have unannotated neighbors to evaluate the robustness of our iterative model. The evaluation result of iterative FSW algorithm is presented in Fig. 7.

Figure 7 shows that 82 out of 100 proteins have identical predicted results from different predicting orders, which demonstrated the robustness of our iterative model. Actually, since the $(t + 1)$th round of prediction only use prediction results from the tth iteration, thus the prediction order in each round of iteration has no influence on the final prediction results.

Table 1. Prediction results of YHL011C. (It can be observed that in three different prediction runs, orders of proteins prediction (for example in 1st iteration of run No. 1, it starts prediction from YNL189W, then YOL061W, finally YHL011C) in each run are different. Predicted functions reached stable status after 3rd iteration in all three runs and different runs achieved the identical predicted functions at last).

	Run No. 1	**Run No. 2**	**Run No. 3**
1st iteration	YNL189W > YOL061W > YHL011C	YOL061W > YHL011C > YNL189W	YHL011C > YNL189W > YOL061W
2nd iteration	YNL189W > YHL011C > YOL061W	YHL011C > YNL189W > YOL061W	YNL189W > YHL011C > YOL061W
3rd iteration	YHL011C > YNL189W > YOL061W	YOL061W > YHL011C > YNL189W	YNL189W > YOL061W > YHL011C
	stable status	stable status	stable status
Predicted functions of YHL011C	01.04 01.03.01.03 02.07	01.04 01.03.01.03 02.07	01.04 01.03.01.03 02.07

The changes in results are mainly caused by randomly selecting function groups as prediction results after the iteration repeats in a pattern. The above experimental results also showed the iterative approach can reach stable status very fast. Actually, predicted functions are selected from the functions of annotated proteins in the prediction domain. The iterative model enlarges the prediction domain of a prediction target protein. Functions in the prediction domain are ranked and selected as the predicted results at the initialization stage. When the iteration starts, due to the incorporation of those unannotated proteins with initially predicted functions into the PPIN, the prediction algorithms take the corresponding interactions and new functions into consideration. So functions are to be re-predicted and re-ranked. With the following iterations, however, the network structure and highly ranked candidate functions remain stable, while changes to the prediction results happen on those lowly ranked functions that endorse each other. On the other hand, the number of function is finite in the prediction domain. Therefore, predicted results will finally reach a stable status after several iterations.

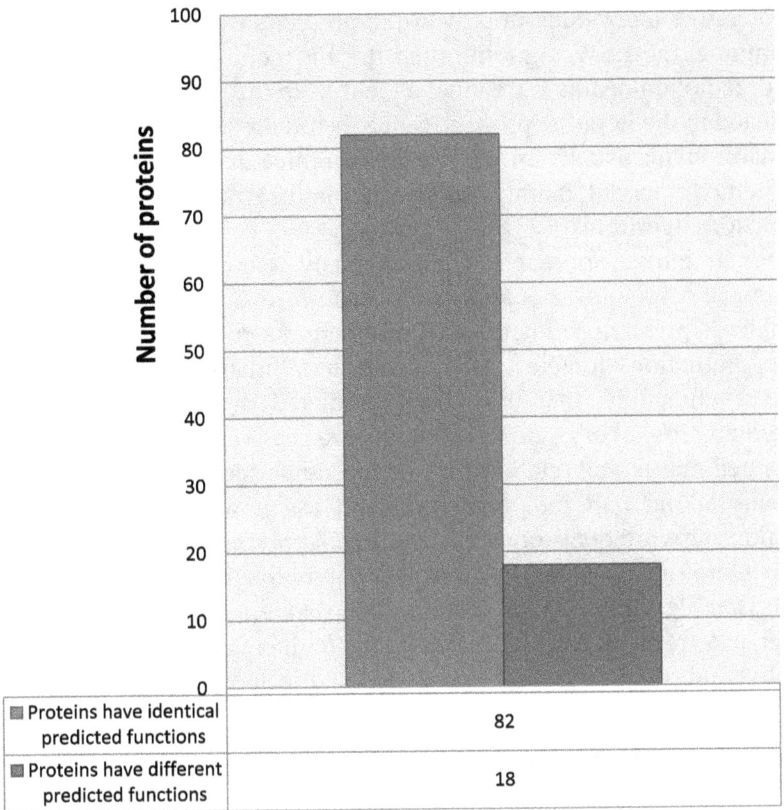

Fig. 7. Numbers of target proteins which had identical predicted functions from multiple independent predictions. (Numbers of the proteins that had identical predicted functions from multiple independent predictions and the proteins that had different predicted functions from multiple independent predictions were presented in this chart. Orders of proteins being predicted were set to be different in each run of prediction).

4. Conclusion

This paper proposes a new iterative model on the PPIN for protein function prediction. Protein interaction information of unannotated proteins in the prediction domain was utilized in the prediction process to improve the prediction results. This model could be applied on existing protein function prediction algorithms and could enhance their prediction performance. In this paper, we employed the NC algorithm, the FSW algorithm and the FF method in our experiments as reference algorithms.

Our iterative algorithms improved the prediction performance of the NC algorithms, the FSW algorithm and the FF method in most cases. Also, most prediction results were independent from the orders of proteins being predicted in the iteration process. Since the predicted functions could reach the stable status usually after no more than three iterations, the time cost of the iterative model mainly depended on the time cost of the chosen prediction algorithms.

Our iterative approach simultaneously recorded functions of all unannotated proteins in prediction domain. So our iterative approach is capable of predicting functions of all unannotated proteins that share the same prediction domain. After iteration, a matrix in which columns indicates functions that have been predicted in each iteration can be generated for every target protein, respectively. Interpreting these generated matrix and relationships between matrices, filtering unannotated proteins taking part into prediction and assigning weights to different iteration steps are questions to be studied. Applying the iterative model on Gene Ontology[34] semantic information based and genomic data based prediction algorithm is our future work. Also, exploiting datasets that are integrated from diverse biological data to increase performance of the iterative approach is going to be conducted in the near future.

Acknowledgment

DW participated in the study of this iterative model for protein function prediction, carried out experiments and result analyses and drafted the manuscript. JH conceived of and coordinated the study, participated in the study of this iterative model, participated in experimental result analyses, revised and finalized the manuscript. All authors have read and approved the final manuscript.

Appendix. Pseudo Code of the Iterative Model

The code mainly consists of four parts, *preliminary*, *protein function initialization*, *iterative function updating* and *prediction results selection*.

Preliminary:
Input PPI data $G(V, E)$, annotation file;
Construct the set of main target protein P_x and other unannotated target proteins in prediction domain as φ;

$P_i \in \varphi$;

$t = iteration\ step$

$FP_{x,t} = set\ of\ functions\ of\ P_x\ from\ t$th iteration;

$FP_{i,t} = record\ of\ predicted\ function\ of\ P_i\ from\ t$th iteration;

$NP_{i,t} = set\ of\ annotated\ proteins\ in\ prediction\ domain\ of\ P_i$;

$F_{\varphi,t} = set\ of\ functions\ in\ \varphi\ after\ the\ t - th\ iteration$, $F_{\varphi,t} = \cup FP_{i,t}$;

$F = set\ of\ all\ functions\ in\ the\ PPI\ network$;

$Frequency(function, protein) = times\ of\ a\ function\ being\ assigned$

$to\ protein \in \varphi$;

$Prediction\ result\ of\ P_x = \emptyset$;

Protein function initialization

For function in F:

    $Frequency(function, P_x) = 0$;

End for

For P_i in φ:

    $FP_{i,1} = predicted\ by\ NC/FSW\ based\ on\ F_{\varphi,0}\ and\ NP_i$;

For function in $FP_{x,1}$:

    $Frequency(function, P_x)+ = 1$

End for

Update NP_i for every P_i;

Update $F_{\varphi,0}$ to $F_{\varphi,1}$ using $FP_{i,1}$;

Iterative function updating:

While not stable:

    For P_i in φ, in the tth iteration:

        $FP_{i,t} = predicted\ by\ NC/FSW\ based\ on\ F_{\varphi,t-1}\ and\ NP_i$;

        For function in $FP_{x,t}$:

            $Frequency(function, P_x)+ = 1$;

    Update $F_{\varphi,t-1}$ to $F_{\varphi,t}$ using $FP_{i,t}$;

    Update NP_i for every P_i;

$t+ = 1$;

End while after reaching stable status after Mth iteration

Prediction results selection:

For function in F:

    If $Frequency\ (function, P_x) \geq \alpha * \max_{f \in U_{1 \leq t \leq M} FP_{x,t}}(Frequency(f, P_x))$:

        Prediction result of $P_x \cup = function$;

Return prediction result of P_x and end for

References

1. Altschul SF, Gish W, Miller W, Myers EW, Lipman DJ, Basic local alignment search tool, *J Mol Biol* **215**:403–410, 1990.
2. Chua NH, Sung WK, Wong L, Predicting protein functions from protein interaction networks, in *Biological Data Mining in Protein Interaction Networks*, IGI Global, Hershey, pp. 203–222, 2009.
3. Schwikowski B, Uetz P, Fields S, A network of interacting proteins in yeast, *Nat Biotechnol* **18**:1257–1261, 2000.
4. Titz B, Schlesner M, Uetz P, What do we learn from high-throughput protein interaction data, *Expert Rev Proteomics* **1**:111–121, 2004.
5. Chua HN, Sung WK, Wong L, Exploiting indirect neighbours and topological weight to predict protein function from protein–protein interactions, *Bioinformatics* **22**(13):1623–1630, 2006.
6. Sharan R, Ulitsky I, Shamir R, Network based protein function prediction, *Mol Syst Biol* **3**:88, 2006.
7. Hishigaki H, Nakai K, Ono T, Tanigami A, Takagi T, Assessment of prediction accuracy of protein function from protein–protein interaction data, *Yeast* **18**:523–531, 2001.
8. Deng M, Zhang K, Mehta S, Chen T, Sun F, Prediction of protein function using protein–protein interaction data, *J Comput Biol* **10**:947–960, 2003.
9. Wu QY, Ye YM, Michael KNg, Ho S-S, Shi R, Collective prediction of protein functions from protein-protein interaction networks, *BMC Bioinformatics* **15**(2):S9, 2014.
10. Xiong W, Xie L, Zhou SG, Guan JH, Active learning for protein function prediction in protein–protein interaction networks, *Neurocomputing* **145**:0925–2312, 2014.
11. Vazquez A, Flammini A, Maritan A, Vespignani A, Global protein function prediction from protein–protein interaction networks, *Nat Biotechnol* **21**:697–700, 2003.
12. Karaoz U, Murali TM, Letovsky S, Zheng Y, Ding C, Cantor CR, Kasif S, Whole-genome annotation by using evidence integration in functional-linkage networks, *Proc Natl Acad Sci* **101**:2888–2893, 2004.
13. Nabieva E, Jim K, Agarwal A, Chazelle B, Singh M, Whole-proteome prediction of protein function via graph-theoretic analysis of interaction maps, *Bioinformatics* **21**(1):i302–i310, 2005.
14. Chi XX, Hou JY, An iterative approach of protein function prediction, *BMC Bioinformatics* **12**:43, 2011.
15. Hou J, Wei Z, Chen YP, Dynamically predicting protein functions from semantic associations of proteins, *Netw Modeling Anal Health Inform Bioinf* **2**(4):175–183, 2013.

16. Wei P *et al.*, Improving protein function prediction using domain and protein complexes in PPI networks, *BMC Syst Biol* **8**:35, 2014.

17. Pizzuti C, Rombo SE, Algorithms and tools for protein–protein interaction networks clustering, with a special focus on population-based stochastic methods, *Bioinformatics* **10**:1343–1352, 2014.

18. Bader GD, Hogue CW, An automated method for finding molecular complexes in large protein interaction networks, *BMC Bioinformatics* **4**:2, 2003.

19. Altaf-Ul-Amin M, Shinbo Y, Mihara K, Kurokawa K, Kanaya S, Development and implementation of an algorithm for detection of protein complexes in large interaction networks, *BMC Bioinformatics* **7**:207, 2006.

20. Adamcsek B, Palla G, Farkas IJ, Derényi I, Vicsek T, CFinder: Locating cliques and overlapping modules in biological networks, *Bioinformatics* **22**:1021–1023, 2006.

21. King AD, Pržulj N, Jurisica I, Protein complex prediction via cost-based clustering, *Bioinformatics* **20**:3013–3020, 2004.

22. Becker E, Robisson B, Chapple CE, Guénoche A, Brun C, Multifunctional proteins revealed by overlapping clustering in protein interaction network, *Bioinformatics* **28**:84–90, 2012.

23. Enright AJ, Dongen SV, Ouzounis CA, An efficient algorithm for large-scale detection of protein families, *Nucleic Acids Res* **30**:1575–1584, 2002.

24. Pereira JB, Enright AJ, Ouzounis CA, Detection of functional modules from protein interaction networks, *Proteins* **54**:49–57, 2004.

25. Ahn Y-Y, Bagrow JP, Lehmann S, Link communities reveal multiscale complexity in networks, *Nature* **466**:761–764, 2010.

26. Pizzuti C, Rombo SE, Restricted neighbourhood search clustering revisited: An evolutionary computation perspective, *Proc 8th IAPR Int Conf Pattern Recognition in Bioinformatics (PRIB)*, pp. 59–68, 2013.

27. Tanay A, Sharan R, Shamir R, Discovering statistically significant bi-clusters in gene expression data, *Bioinformatics* **18**(1):S136–S144, 2002.

28. Lan L, Djuric N, Guo Y, Vucetic S, MS-kNN: Protein function prediction by integrating multiple data sources, *BMC Bioinformatics* **14**(3):S8, 2013.

29. Yijie W, Xiaoning Q, Joint clustering of protein interaction networks through Markov random walk, *BMC Syst Biol* **8**(1):S9, 2014.

30. Chua HN, Sung WK, Wong L, An efficient strategy for extensive integration of diverse biological data for protein function prediction, *Bioinformatics* **23**(24):3364–3373, 2007.

31. Ashburner *et al.*, Gene ontology: Tool for the unification of biology, *Nat Genet* **25**(1):25–29, 2000.

32. Ito T, Tashiro K, Muta S, Ozawa R, Chiba T, Nishizawa M, Yamamoto K, Kuhara S, Sakaki Y, Toward a protein–protein interaction map of the budding yeast: A comprehensive system to examine two-hybrid interactions

in all possible combinations between the yeast proteins, *Proc Natl Acad Sci* **97**:1143–1147, 2000.

33. Ito T, Chiba T, Ozawa R, Yoshida M, Hattori M, Sakaki Y, A comprehensive two hybrid analysis to explore the yeast protein interactome, *Proc Natl Acad Sci* **98**:4569–4574, 2001.
34. Ashburner M *et al.*, Gene ontology: Tool for the unification of biology, *Nat Genet* **25**(1):25–29, 2000.
35. Obayashi T, Kinoshita K, COXPRESdb: A database to compare gene coexpression in seven model animals, *Nucleic Acids Res* **39**:D1016–D1022, 2011.
36. Ruepp A *et al.*, The FunCat, a functional annotation scheme for systematic classification of proteins from whole genomes, *Nucleic Acids Res* **32**(18): 5539–5545, 2004.

Derui Wang received his Bachelor degree of Engineering in Optoelectronic Information Engineering from Huazhong University of Science and Technology (HUST), China, in 2011. Currently, Derui is a M.Sc. by research student in Deakin University, Australia. His research interests include machine learning, data mining and bioinformatics.

Jingyu Hou received his first Ph.D. degree in Computational Mathematics from Shanghai University, China (1995) and the second Ph.D. degree in Computer Science from the University of Southern Queensland, Australia (2004). Jingyu is currently with the School of Information Technology, Deakin University, Australia. His research interests include bioinformatics, data and Web mining, databases and information retrieval and Web engineering.